Blogging Wildlife

This volume reports on the encounters between hikers and wildlife on the Appalachian Trail. Based on narratives provided by trail hikers, it explores the ways in which humans relate to the animals with whom they temporarily share a home. With attention to the themes of pilgrimage, the changing perception of the animals encountered and reactions to them, risk, auditory experience, and a sense of wildness, the author considers the meaning constituted by non-human animals in the context of the walkers' narrative journeys. A phenomenologically informed study of the ways in which people perceive wild animals when in an unmediated wilderness setting, how they navigate interactions with them, and how they experience living among them, *Blogging Wildlife* will appeal to scholars across the social sciences with interests in anthrozoology and human–animal relations.

Kate Marx was a postdoctoral research associate in the research group Exeter Anthrozoology as Symbiotic Ethics (EASE) at the University of Exeter, UK. She is currently Social Research Manager at Waterwise, a not-for-profit organisation which advocates for water conservation in the UK.

Multispecies Encounters

Series editors:

Samantha Hurn is Associate Professor in Anthropology, Director of the Exeter Anthrozoology as Symbiotic Ethics (EASE) working group and Programme Director for the MA and PhD programmes in Anthrozoology at the University of Exeter, UK.

Chris Wilbert is Senior Lecturer in Tourism and Geography at the Lord Ashcroft International Business School at Anglia Ruskin University, UK.

Multispecies Encounters provides an interdisciplinary forum for the discussion, development, and dissemination of research focused on encounters between members of different species. Re-evaluating our human relationships with other-than-human beings through an interrogation of the 'myth of human exceptionalism' which has structured (and limited) social thought for so long, the series presents work including multi-species ethnography, animal geographies, and more-than-human approaches to research, in order not only better to understand the human condition, but also to situate us holistically, as human animals, within the global ecosystems we share with countless other living beings.

As such, the series expresses a commitment to the importance of giving balanced consideration to the experiences of all social actors involved in any given social interaction, with work advancing our theoretical knowledge and understanding of multi-species encounters and, where possible, exploring analytical frameworks which include ways or kinds of 'being' other than the human.

Published

Animal Places
Lively Cartographies of Human–Animal Relations
Jacob Bull, Tora Holmberg, and Cecilia Åsberg

Human–Canine Collaboration in Care
Doing Diabetes
Fenella Eason

Living-With Wisdom
Permaculture and Symbiotic Ethics
Alexander Badman-King

Blogging Wildlife
The Perception of Animals by Hikers on the Appalachian Trail
Kate Marx

The full list of titles for this series can be found here: www.routledge.com/Multispecies-Encounters/book-series/ASHSER1436

Blogging Wildlife

The Perception of Animals by Hikers
on the Appalachian Trail

Kate Marx

Routledge
Taylor & Francis Group

LONDON AND NEW YORK

First published 2021
by Routledge
2 Park Square, Milton Park, Abingdon, Oxon OX14 4RN

and by Routledge
52 Vanderbilt Avenue, New York, NY 10017

Routledge is an imprint of the Taylor & Francis Group, an informa business

British Library Cataloguing-in-Publication Data
A catalogue record for this book is available from the British Library

Library of Congress Cataloging-in-Publication Data
Names: Marx, Kate, 1981– author.
Title: Blogging wildlife: the perception of animals by hikers on the
Appalachian Trail / Kate Marx.
Description: Abingdon, Oxon; New York, NY: Routledge, 2021. |
Series: Multispecies encounters | Includes bibliographical references and index.
Identifiers: LCCN 2020039837 (print) | LCCN 2020039838 (ebook) |
ISBN 9780367351007 (hardback) | ISBN 9780429329722 (ebook)
Subjects: LCSH: Human-animal relationships–Appalachian Trail.
Classification: LCC QL85 .M3136 2021 (print) |
LCC QL85 (ebook) | DDC 590.974–dc23
LC record available at https://lccn.loc.gov/2020039837
LC ebook record available at https://lccn.loc.gov/2020039838

ISBN: 9780367351007 (hbk)
ISBN: 9780429329722 (ebk)

Typeset in Times New Roman
by Newgen Publishing UK

Contents

Acknowledgements

The chapters in this book draw upon material in the publications listed below. The author is grateful to the publishers for their permission to include selected portions and adaptations of this material in the book.

Marx, K. 2019. "He's so Fluffy I'm Gonna Die!" Cute Responses by Hikers to Autonomous Animals on the Appalachian Trail. *Anthrozoös* 32(1): 89–101. Copyright © International Society for Anthrozoology, reprinted by permission of Taylor & Francis Ltd, http://tandfonline.com on behalf of the International Society for Anthrozoology.

Marx, K. 2019. Transgressive Little Pests: Hiker Descriptions of "Shelter Mice" on the Appalachian Trail. *Anthrozoös* 32(1): 103–115. Copyright © International Society for Anthrozoology, reprinted by permission of Taylor & Francis Ltd, http://tandfonline.com on behalf of the International Society for Anthrozoology.

Marx, K. 2018. Hiker Narratives of Living among Bears on the Appalachian Trail. *Society & Animals* 1(10 Dec): 1–19. Copyright © Brill, reprinted by permission.

The Trek, 2017. Extracts © The Trek, reprinted by permission of the Trek. Available at: thetrek.co

Introduction

For people who live in contemporary Western societies, engagement with autonomous nonhuman animals can be limited or close to non-existent (see Berger, 1980; Bulbeck, 2005; Bulliet, 2005; Malamud, 1998; Turner 1996). Consequently, ideas about 'wild' animals are formed primarily from cultural sources (which, from bedtime stories to nature documentaries, can only ever offer a pale representation of the 'real' animal) or from zoos or organised wildlife encounters during which human and animal interactions are heavily managed (see Anderson, 1998; Berger, 1980; Bulbeck, 2005; Curtin, 2005; Franklin, 1999; Keul, 2013; Malamud, 1998; McNamara and Prideaux, 2011; Rothfels, 2002; Schänzel and McIntosh, 2000). For most of us, our experience of autonomous animals is shaped almost entirely by cultural artefacts: plush toys, Disney films, wildlife documentaries, animal tattoos, 30-second comedy clips posted on social media platforms, charity appeals, taxidermy museum exhibits, and breakfast cereal packaging. The notion that our media culture controls and shapes perceptions of wild animals has been extensively argued (see Baker, 2001; Berger, 1980; Bulliet, 2005; Franklin, 1999). Ingold questions whether unmediated perception of the animal is even possible for us:

> Do animals exist for us as meaningful entities only insofar as each may be thought to manifest or exemplify an ideal type constituted within the set of symbolic values making up the 'folk taxonomy' specific to our culture? Or do we perceive animals directly, by virtue of their immersion in an environment that is largely ours as well, regardless of the images that we may hold of them, or of whether we hold such images at all?
>
> (Ingold, 1994: 12)

For many people, there is such little opportunity to perceive (wild) animals directly, that the folk taxonomy that Ingold refers to is really all they have to go on when thinking about certain animals, and the question, with regard to autonomous animals at least, goes unanswered.

Over recent years there has been an increasing sense of impoverishment associated with our lack of contact with self-willed animals (see Bulbeck, 2005; Turner, 1996). This has been accompanied by a rise in modes of tourism focused on the immersion of the tourist in 'natural' settings (nature tourism), the experience of nature with a conservation element (ecotourism) and tourist encounters with wild animals (wildlife tourism). These modes of tourism often overlap each other, but all are ultimately concerned with giving the tourist an encounter with 'authentic wildness' (see Bulbeck, 2005; Cater, 2006; Curtin, 2005; Dearden, 1990; Fullagar, 2000; Hill et al., 2014; Keul, 2013; Kubo and Shoji, 2016; McNamara and Prideaux, 2011; Orams, 2000, 2002; Reynolds and Braithwaite, 2001; Schänzel and McIntosh, 2000; West and Carrier, 2004). The engagement with wild animals that this type of tourism offers is often tightly managed, featuring boundaries, fences, boardwalks, site maps, viewing times, feeding times, guides, rangers, crowds of other visitors, and educational lectures, as well as strict regulations about what the tourist can do with her body, frequently including rules such as no stopping, no lingering, no approaching the animals, no reaching out towards the animals, no touching, no feeding, no prolonged eye contact. The tourist's perception of, and engagement with, the animals featured in an encounter like this is unavoidably shaped by these multiple forms of mediation, which interrupt any possibility of being able to perceive the animal 'directly'.

There *are* ways for people to engage with autonomous animals in less managed settings. One way is for us to take notice of all the wildlife living around us already – the birds, insects, amphibians, small mammals – whom we rarely seem to interact with, despite often living in close proximity to each other. A more extreme way is to go and live in the spaces belonging to wild animals, referred to (somewhat controversially) by some as the 'wilderness', a word used to describe places where humans and the visible artefacts of human society are outnumbered by the self-willed nonhuman animals living there, and where humans must navigate encounters with these animals, and find ways of co-existing with a multitude of species (see Nash, 1982; Oelschlaeger, 1991; Salleh, 1996). The increasing popularity of wilderness recreation – in particular, long-distance hiking – means that more and more people are entering the wild homes of autonomous animals and experiencing face-to-face encounters with them for the first time (see The Trek, 2017).

This book explores how people perceive autonomous nonhuman animals when in a so-called wilderness setting, how they navigate interactions with them, how they experience dwelling among them, and what kinds of stories they tell themselves (and others) about them. The research focuses on a cohort of long-distance hikers who trekked all or part of the Appalachian Trail (AT), a more than 2,000-mile-long wilderness trail in the United States, during the years 2015 and 2016. For many, perhaps most, of these hikers, this was their first experience of engaging with wild animals on an 'equal footing', with no intermediary guiding the experience and no fences keeping them or the animal in place. It was clear from hiker narratives that they and the autonomous animals on the trail affected each other. Perceptions of animals are far from homogenous. When encountering certain species some hikers described carefully backtracking and finding an alternate path in order not to disturb them, while others talked about catching, killing, and eating them. Because the way we perceive autonomous animals also affects how we behave towards them, these perceptions deserve close attention.

By taking a phenomenological approach to people's perception and experiences of self-willed animals, looking at the influences affecting how people perceive autonomous animals, and how they incorporate their experiences into their lifeworlds, we see that our experiences of wild animals are not an either/or scenario (*either* we see them through the prism of our cultural knowledge about them *or* we experience them directly through our own sensory contact with them), but instead, are formed through a complex combination of factors, including their cultural construction or position in the existing folk taxonomy (see Baker, 2001; Berger and Luckmann, 1991; Evernden, 1992; Hurn, 2012; Ingold, 1994), allegedly innate or learned reactions like disgust or the 'cute response' (see Dale et al., 2017; Lorenz, 1971; Ngai, 2012), the somatic and sensory aspects of the specific encounter (see Abram, 1997; Almagor, 1990; Buller, 2012; Keul, 2013; Milton, 2005), environmental factors such as the space in which the encounter takes place (see Foucault and Miskowiec, 1986; Ingold, 2011; Tuan, 1977; Ulrich, 1983), the context and meaning that people attribute to their encounter, why they are on the trail to begin with (see Badone and Roseman, 2004; Lyng, 1990; Schänzel and McIntosh, 2010; Urry, 1990), and the identity-building work that people can use wild animal encounters to carry out (see Azariah, 2016; Bourdieu, 1986; Fullagar, 2000).

There is a tendency for humans in Western societies to view ourselves as the keepers of *all* nonhuman animals, something which interacting with free animals in unmanaged settings has the potential to disrupt – to our benefit, if the argument that contact with wild animals is vital for our own mental health is to be believed (see Bulbeck, 2005; Leopold, 1968 [1949]; Milton, 2002; Thoreau, 1995 [1854]; Turner, 1996).

The research presented here centres around blog posts written by people who undertook a long distance hike of the AT in recent years. The blogs are a mainline into the narrator's own perception of their experience as well as into the way in which the narrator chooses to frame and make sense of their experience in the form of a story. The voices of the individual bloggers, talking about dwelling among and interacting with many different wild species, are lively and distinctive. Significantly, they also offer an insight into what the other-than-human animal may have experienced from the encounter. In his *Phenomenology of Perception,* Merleau-Ponty describes the way in which another person's experience can be 'taken up' into our own:

> The phenomenological world is not pure being, but rather the sense that shines forth at the intersection of my experiences and at the intersection of my experiences with others through a sort of gearing into each other.
>
> (Merleau-Ponty, 2014 [1945]: lxxxiv)

Through unmanaged contact with autonomous animals, hikers came to incorporate the animals into their lifeworlds. This assimilation into the lifeworld of the hiker can potentially be viewed as the animal *becoming interesting* to the hiker. Despret (2011) has described becoming interesting (in the eyes of the human) as a political achievement for animals, suggesting that when animals are interesting to us, it creates the opportunity for a new, more understanding, relationship with them. At the same time, however, it is clear that hikers never fully moved away from the idea of the animal *meaning something*. The autonomous animals on the trail were experienced in complex, multifaceted, sometimes even contradictory ways that succeeded in making them interesting to hikers in ways that they hadn't anticipated, and yet hikers also thought about animals as meaning, or representing, something, in the context of their own narrative journey.

The trail

Running from Springer Mountain, Georgia in the south to Mount Katahdin, Maine, in the north, the AT in North America traverses 14 US states, covering roughly 2,200 miles of forest, mountains, ridges, and plains (Appalachian Trail Conservancy (ATC), 2016: np) as well as passing through farmland, about a dozen small towns, and a zoo. It is the world's longest hiking-only trail, and every year millions of people from across the United States, and many from across the world, travel to hike it. Many plan to walk a section of it and a minority intend to cover the entire length of the trail, estimated to take between five and seven months to complete. On average, one in four hikers attempting the 'thru-hike' makes it to the end (ATC, 2016: np).

The AT evolved from a 1921 proposal by a regional planner named Benton MacKaye, who conceived of it essentially as a series of campgrounds connected by a hiking trail, which Eastern US city dwellers could escape to for recreation and recuperation for a couple of weeks each year. MacKaye (1921: 325) wrote that "the ability to cope with nature directly – unshielded by the weakening wall of civilization – is one of the admitted needs of modern times". Construction was carried out in the 1920s and 1930s by teams of volunteer hiking clubs (NPS, 2008).

Depending on what section of the trail a hiker is on, they can encounter wildlife such as deer (*Odocoileus virginianus*), elk (*Cervus canadensis*), beavers (*Castor canadensis*), moose (*Alces alces*), black bears (*Ursus americanus*), feral ponies (*Equus ferus caballus*), porcupines (*Erethizon dorsatum*), salamanders (*Caudata*), turkeys (*Meleagris gallopavo*), foxes (*Vulpes vulpes*), rattlesnakes (*Crotalus horridus*), copperhead snakes (*Agkistrodon contortrix*), bald eagles (*Haliaeetus leucocephalus*), and numerous other species. A plaque at Springer Mountain in Georgia describes the trail as "a footpath for those who seek fellowship with the wilderness" (ATC), 2016: np), and the ATC, the organisation charged with the (often conflicting) jobs of promoting and protecting the AT, describe the trail as "intentionally built and maintained in a manner in keeping with wilderness values" (ATC, 2016: np), citing inconspicuous signage and the fact that trail maintainers are required to use hand tools to clear vegetation, as bringing power tools onto the trail would conflict with the wilderness ethic. In fact, the 'wilderness' of the AT is in a constant state of flux. Much of the trail has, in effect, been rewilded, with new forest allowed to take over where farming or logging once took place. Between the protected areas that it passes through – federal, state, and local parks and forests – the AT runs through a corridor of land sometimes no more than 1,000 feet in width. Hikers, organisations such as the ATC and the National Park Service (NPS), volunteer trail maintenance crews (who brush out the trail, clear fallen trees from it, and paint the white blazes that mark the route), trail-runners (who are allocated a section of trail to work, informing and educating hikers about 'Leave No Trace' principles), phenology and wildlife monitoring volunteers, journalists and authors, and the local other-than-human animals themselves all collaborate in the physical and social construction of the AT 'wilderness'.

Over nearly a century that the AT has existed, the focus of its stewards (the ATC, the NPS, teams of trail maintenance volunteers) has shifted significantly from maintaining the trail for the use and enjoyment of hikers, to "conserving and managing the significant natural and cultural resources that make the trail such an exceptional resource" (Shriver et al., 2005: np). Extensive brochures like the *Appalachian National Scenic Trail Resource Management Plan* (NPS, 2008) discuss how to monitor, manage, and protect the 'biological resources' (flora and fauna), 'air resources', 'water resources', and 'cultural resources' (archaeological remains, sites of sacred significance to Native Americans, and man-made elements of the trail valued by hikers, for example some famous shelters) of the trail; an increasingly challenging task in the face of local development pressures and the greater numbers of hikers using the trail every year. The goals detailed in the *Appalachian Trail Conservancy Strategic Plan* (ATC, 2014) include both strategies for dealing with the negative impact of hikers on the trail, and strategies for attracting more hikers to the trail.

Why hike the AT?

The number of people thru-hiking (trekking from one end to the other) the AT has risen dramatically in recent years. According to the ATC, between 1936 and 1969 only 59 thru-hike completions were recorded (2016: np). In 1970, ten people completed the trail. By 1980, this number had increased more than ten-fold. It had doubled by 1990 and again by 2000. The ATC (2016: np) states that "more hike completions were recorded for the year 2000 alone than in the first 40 years combined". To date, more than 15,000 thru-hike completions have been recorded by the ATC, representing only around a quarter of the people who set out with the intention of completing the trail. The many books written by former thru-hikers about their time on the trail (including Bryson, 1998; Cornelius, 1991; Davis, 2012; and Hall, 2000) have undoubtedly publicised it as a destination, but none of them make the trail sound easy, or pleasant. Indeed, they emphasise the difficulty and deprivation that's involved in a thru-hike. Yet every year more people set out for their own six-month 'walk in the woods'. Why?

In her ethnography of the AT, Fondren notes that "an overwhelming majority of long distance hikers are at a transitional stage in life and searching for deeper meaning" (2016: 32). Many talk about leaving university and not wanting to be funnelled into a nine-to-five job just yet, or having worked 20 years in an office and feeling like there must be more to life (see The Trek, 2016). One of Klein's interviewees decided to thru-hike in honour of his dog, who had recently died (2015: 129). Not all, but many, hikers cite a general disillusionment with modern life, with what they see as the banalities of contemporary society: bosses, bills, binge-watching Netflix. Fondren also notes that many long-distance hikers on the AT are there because they want to escape from modern life, saying that "the Appalachian Trail is valued for its ability to provide recreationists with opportunities for social interaction and deep connections to place" (2016: 11). Goldenberg et al. (2008) found that people's reasons for hiking the AT included undertaking a physical challenge, learning survival skills and self-reliance, increasing self-esteem and self-fulfilment, solitude, camaraderie, and companionship (solitude and companionship are both available on the AT; it is possible to hike the entire length on your own, and also possible to join and remain in the hiker 'bubble' – the group of hikers that set off at the most popular time of year and stay more or less together for the entire length of the trail. In reality, many hikers choose a combination of both; it is not frowned upon to hike with companions for weeks at a time and then leave early one morning to continue the hike on your own). They concluded that the main reasons for people to hike the AT were for "fun and enjoyment of life, and warm relationships with others" (Goldenberg et al., 2008: 279). Fondren's description of AT thru-hikers as "recreationists" is not entirely convincing, nor is Goldenberg's conclusion that most hikers are there for "fun". An AT thru-hike is an enormous commitment. The ATC estimates that it costs between $1,000 – $2,000 (around £800 – £1,600) to buy the gear necessary to hike, and that on average hikers can expect to spend a further $1,000 per month on their hike (ATC, 2016: np). For an average six month thru-hike this works out to between $7,000 and $8,000 (around £5,500 to £6,500). In addition, for their six months on the trail the AT hiker may not have any income to speak of; indeed, many talk about having to quit their jobs in order to hike (The Trek, 2016: np). They also often leave behind mortgages or leases on homes, spouses, children, friends, and all the routines and familiarity of the lives that they are used to. Some sell their cars and/or most of their belongings. More than once Trek hiker-bloggers wrote about missing the birth of a grandchild while on the trail. One man had a brain tumour and had been given two years left to live by his doctor. He had decided he wanted to spend at least the next six months of the remainder of his life hiking the AT (The Trek, 2016: np). Bratton (2012: 152) interviewed one AT hiker whose wife and child had died in an automobile accident, and whose father had "put him on the trail" to help him to cope with his grief. In short, the commitment and sacrifices that many people make to spend their time hiking a trail through the woods speaks to a motivation more complex and nuanced than just 'fun'.

In Goldenberg et al.'s study, "being outdoors" and "environmental awareness" (2008: 279) were mentioned, but do not appear to have been as significant as the other factors cited. However, many of the participants that Fondren interviewed did mention wanting to connect with 'nature' or 'the wilderness'. One participant said that there were "a lot of people out here seeking a spiritual, close-to-nature thing" (2016: 67). For many of Fondren's participants (as well as Bratton's, 2012), a thru-hike was likened to a religious pilgrimage, with one equating being in the woods to being in a cathedral. Throughout hiker blogs it becomes clear that the woods, nature, and the wilderness are equated with the more abstract ideals that people hoped to find while on the AT, particularly 'freedom', 'authenticity', and 'connection'; themes which greatly affect how the nonhuman animals on the trail were experienced and perceived by hikers.

A wild life and wildlife on the trail

In contrast to the numerous studies on people who travel to wildlife destinations specifically in order to engage with local animal life (for example, Bulbeck, 2005; Curtin, 2005; Duffus and Dearden, 1990; Keul, 2013; Orams, 2000, 2002; Reynolds and Braithwaite, 2001; Russell, 1995; Schänzel and McIntosh, 2010), it is important to emphasise that not everybody who hikes the AT does so in order to encounter, or interact with, wildlife. Rather, there is frequently a more general sense of wanting to be in 'wilderness'. But what really *is* wilderness? In Old English the word "willed" (meaning self-willed, autonomous, uncontrollable) and the word "deor" (meaning deer specifically, but also animals in general) were combined to form "wild-deor-ness" – "the place of wild beasts" (Nash, 1982: 2). Wilderness literally refers to the place where wild animals are, and thus it is the animals who make the wilderness. It is the wild, self-willed animals that make the AT the wilderness that so many people quit their jobs, spend their savings, and leave their loved ones to live in. It is these animals that are inseparable from the ideals that many hikers associated with living on the AT: freedom, authenticity, connection. Yet, rather than experiencing a desire to interact with animals, per se, for hikers the emphasis was more on the idea of *being with them*. Bulbeck (2005: 149–150) speaks of a trip to Antarctica, saying, "my fellow expeditioners…were awed by their experience rather than connected with other creatures; we were in communion ('fellowship') rather than in communication ('conversation') with nature". It was this sense of "communion" that was sought by many, rather than, as Bulbeck puts it, "communication", and therefore, when talking about people's 'experience' of an animal, it will often be in reference to their experience of communion or dwelling with the animal, as opposed to face-to-face contact.

Studies centred around the AT

The AT has been of increasing interest to researchers from several different disciplinary backgrounds, and is the focus of numerous studies. To start, the inception and construction of the trail has been covered from a historical perspective (see Kates, 2013; Mittlefehldt, 2010, 2013). As discussed earlier, social scientists have examined why people choose to hike the trail (see Fondren, 2016; Goldenberg et al., 2008; Gomez et al., 2010; Hill, 2014), and have also explored the physical and mental benefits of undertaking a hike (see Bratton, 2012; Freidt et al., 2010; Hill, 2014; Hill et al., 2009). Attempts have been made at understanding how hikers use the trail, including quantifying hiker footfall (Zarnoch et al., 2011) and evaluating how knowledgeable they are about minimum impact practices (Newman et al., 2003). Fondren's 2016 ethnography of AT thru-hikers explores the experiences of long-distance hikers from several different angles, while other studies focus specifically on the spiritual elements of a thru-hike (Bratton, 2012), conflict between thru-hikers (MacLennan and Moore, 2011), the relationship that hikers have with their equipment (Littlefield and Siudzinsk, 2012), and hiking as performance (Terry and Vartabedian, 2013). The importance of place, and hiker relationships to the environment of the AT, have also been the focus of several studies (see Dorwart et al., 2009; Gerard et al., 2003; Klein, 2015; Kyle et al., 2010). Research looking at the experiences on the trail of specific groups, including military personnel in training (Smith et al., 2013), women (Boulware, 2004), and individuals with specialised needs (Nisbett and Hinton, 2005) has also been undertaken.

In terms of the interaction between hikers and animals on the trail, there have been a few studies conducted. The presence on the trail of organisms that can be detrimental to hiker health has been quantified, for example, bacteria in water sources (Reed and Rasnake, 2016) and ticks (Ford et al., 2015), while anthropogenic influences on mammal occupancy along the AT have also been looked at (Erb et al., 2012). Although not specifically focused on the AT, researchers have been particularly interested in the impact of black bears (*Ursus americanus*) and campers upon each other in the parks that the AT runs through (see Baptiste et al., 1979; Burghardt et al., 1972; Clark et al., 2003; Gore et al., 2006; Pelton, 1972; Pelton et al., 1976; Singer and Bratton, 1980; Tate and Pelton, 1983). These bear-focused studies will be discussed in Chapter 2.

The Trek blogosphere

The Trek (TheTrek.co) is a long-distance hiking-dedicated website, inspired by the book *Appalachian Trials* (Davis, 2012), written by former thru-hiker

Zach Davis. The website features advice on all aspects of long-distance hiking the AT (and other long-distance trails, including the Pacific Crest Trail and the John Muir Trail). Segments include gear reviews, articles on thru-hiking culture, and motivational messages. However, the dominant feature of the Trek site is its community of bloggers, many of whom are either planning to thru-hike the AT, are currently thru-hiking, have successfully completed their thru-hike, or have 'dropped out' of their attempted thru-hike.

The vast majority of blog posts are written by hikers currently in the process of their attempted thru-hikes, and are written in the format of narratives about their experiences while on the trail. These narratives tend to focus on certain aspects of the experience: the people that they meet along the way, the nonhuman animals that they encounter, the scenery, the weather, the physical toll on their bodies of hiking day after day carrying a heavy pack, the constant hunger and consequent obsession with food, and the mental challenge of spending months walking through the woods. The master narrative is one of almost relentless physical and mental suffering, interspersed with moments of sublime clarity that make it 'all worth it'. Whether or not they make it to the end of the trail, many hiker-bloggers' narratives report a deep connection to place brought on by, for example, a beautiful view, an encounter with a wild animal, or a campfire conversation with a new friend. The thru-hikers who successfully finish, or report successfully finishing the trail, compose satisfyingly complete narratives of triumph against adversity.

Method

Anthrozoology, sometimes called human–animal studies, is an emergent, interdisciplinary approach to scholarly research which aims to understand and document human interactions with, and perceptions of, nonhuman animals, and specifically "the social, political, economic or historical circumstances which have led to such interactions" (Hurn, 2010: 27). In order to achieve this, anthrozoology utilises the theoretical and methodological approaches of a number of disciplines that have historically concerned themselves with the topic of human–animal interactions; primarily anthropology (e.g. Bird-David, 1999; Candea, 2010; Caplan, 2010), sociology (e.g. Arluke and Sanders, 1996; Bulbeck, 2005; Franklin, 1999), cultural geography (e.g. Anderson, 1998; Buller, 2012; Wolch and Emel, 1998), and history (e.g. Isenberg, 2002; Rothfels, 2002), while also intersecting with cultural studies (e.g. Baker, 2001; Malamud, 1998), media studies (e.g. Gall Myrick, 2015; Golbeck, 2011; Laforteza, 2014), psychology (e.g. LoBue and DeLoache, 2008; Mulkens et al., 1996; Öhman and Mineka, 2003), philosophy (e.g. Despret, 2011; van Dooren, 2014), biology (Isbell, 2006; Wilson, 2005), ethology (e.g. Archer and Monton, 2010; Lorenz, 1971; McArthur Jope, 1983), and more. Indeed, Hurn suggests that "anthrozoology" should refer to any research, regardless of the researcher's disciplinary orientation, "which documents human interactions with other animals from the exclusively human perspective" (2010: 28).

Telling stories: doing narrative research

The aim of this research was to explore (a) what happened between hikers and wildlife on the AT; (b) how hikers thought and felt about the animals they encountered; and (c) how hikers subsequently constructed their experiences with other animals into narratives for the consumption of their readers. Were people on the trail *interested* in the animals that they met? To what extent were their experiences of animals mediated by the cultural representations of wildlife that they had been exposed to? To what extent did their embodied interactions with animals leave them with a new kind of knowledge based on direct experience rather than on pre-existing 'folk taxonomies'? Did they see personhood in the animals on the trail, or did they see something else? Were their encounters transformative in some way, or did they not live up to expectations? What was the experience of interacting with a diverse array of wild species like for hikers (including what it was like for them to think about and reconstruct afterwards), and *why* was it like that?

The research in this book is based wholly around narratives posted by AT hikers onto the Trek website. A total of 1,691 blog posts, written and uploaded to the Trek (the trek.co) during the years 2015 and 2016, by 166 hiker-bloggers, were reviewed to find references to encounters between hikers and wildlife on the trail. The blog posts naturally lent themselves to a narrative research approach, and once the collection of narratives that identified encounters between hikers and wildlife had been assembled, they were thematically coded, using intuitive analysis to seek out commonalities (and divergences) between narratives.

Sugiyama (2005: 178) writes that "storytelling is the product of a mind adapted to hunter-gatherer conditions, and it emerged when our ancestors practiced a foraging way of life". In other words, stories were invented so that people could tell each other about the food source they had just seen on the other side of the mountain, or warn each other of the predator lurking in the woods. Hikers who tell stories about a bear 'stalking' a particular campsite on the trail, or about stumbling upon a previously unknown water source in the woods were returning to an intriguingly similar format to the oldest stories ever told.

Sugiyama (2005: 185) cites Black and Bower's (1980) description of the "essence" of storytelling as a description of a problem and the character's plans for solving it, while Brubaker and Wright similarly describe narrative as a "tool individuals use to deal with disruptions in their lives" (2006: 1217). Besnier refers to Bruner's proposal that through stories we construct a self that is able to meet the challenges of a situation through our memories of the past and hopes/fears for the future, leading Besnier to conclude that storytelling is therefore "a temporal articulation of the past with the future through the present" (2016: 89). When looking at people's AT animal-encounter stories we find out not just what happened during the encounter, but how the experience shaped the person's perception of the animal, which is no doubt influenced by all kinds of factors, including the hiker's pre-existing, culture-based ideas about that animal (their past) as well as their possible intentions regarding the effect of their story on its consumers (cultural capital, a job as a journalist etc. – their future), intersected with the event itself to make a story.

Constructing stories, making lives

In looking to the past for guidance on how to deal with the present, hikers – most of whom will not have had to deal with a charging bear or a rattling snake before – will often have only other people's stories to refer to. In this way, as Jerome Bruner puts it, "Narrative imitates life, life imitates narrative" (2004: 692). Indeed, he says that "the ways of telling and the ways of conceptualizing that go with them become so habitual that they finally become recipes for structuring experience itself...a life as led is inseparable from a life as told..." (2004: 708). In blogger narratives people's recorded responses to certain animals tended to conform to certain types of acknowledged options for dealing with that particular type of animal ('predator', 'pest', etc.), where the full facts of the situation may well be a lot messier and less satisfyingly structured than those chronicled in the narrative. In "What If Foucault Had Had A Blog" (2012), Zylinska states that

> Foucault associates the practice of self-writing precisely with an ethos of life when he claims that 'writing transforms the thing seen or heard into tissue and blood'. From this perspective diaries and blogs are not just commentaries *on* someone's life but materializations *of* it.
>
> (Zylinska, 2012: 62)

It is perhaps possible to rephrase Bruner's quote, to say that myth informs life, and life becomes subsumed back into myth. In people's adventures on the AT the myth of wilderness adventure is so pervasive that in some ways it is almost inevitable that their encounters with wild animals be experienced and recreated through that tradition. Bruner expands upon Goodman's description of painting or history as "world making" with the idea of autobiography as "life making" (2004: 692), concluding that in storytelling it is therefore the form, and not the content, that matters most. He says that, "in the end, we *become* the autobiographical narratives by which we 'tell about' our lives...we also become variants of the culture's canonical forms" (2004: 694). As Berger and Luckmann put it, "men [sic] must talk about themselves until they know themselves" (1991: 53). In this way it is possible to look at a hiker's decision to write about her hike as a means of reinforcing a particular way that she would like to see herself (the adventurer, the nonconformist etc.), of almost willing that version of herself into being. Indeed, when telling stories about ourselves, as opposed to about other people or even fictional characters, our choices about the telling become even more complex because we are, consciously or unconsciously, trying to achieve aims other than a straightforward record of events. Self-telling is an act of self-making (see Besnier, 2016; Brubaker and Wright, 2006; Bruner, 2004; Schiffrin, 2009; Trainer et al., 2016; Wilson, 2005; Zylinska, 2012), and also a way of re-making messy events into something coherent, not only for our audience but also for ourselves. Yet, while autobiographical narrative is used for solving dilemmas, it also introduces problems of its own, for example the need for a particular form of reflexivity which can negotiate the tensions created by the author also playing the leading role in a story about themselves (see Bruner, 2004: 693).

The common knowledge is that there are no new stories. Perhaps part of the problem in avoiding the distortion of what the narrator wants to say is the difficulty of telling a story that does not immediately fit into the canon of what we already know. Wilson (2005: ix) describes the mind as "a narrative machine, guided unconsciously by the epigenetic rules in creating scenarios and creating options. The narratives and artefacts that prove most innately satisfying spread and become culture". It's possible to surmise, then, that in telling a story we instinctively reach for the most narratively satisfying elements of our story and emphasise those, thereby distorting what we actually experienced in deference to a predetermined master narrative, a template that looks and sounds similar to what we experienced, even if only superficially. Deborah Schiffrin (2009: 422) suggests that master narrative could potentially be seen, "such that it is a social and/or cultural resource for organizing communal ideologies and socializing members of a community". In telling stories, the narrator understands what is expected of them narratively, and structures their story accordingly. The AT hiker, blogging in a community of fellow AT hikers, narrates experiences with animals that will help to further embed them in the AT community. Bernard (1995: 147), speaking on the subject of conducting social research, states that "the key to understanding the culture of loggers, lawyers, bureaucrats, schoolteachers, or ethnic groups…is to become intimately familiar with their vocabulary. Words are where the cultural action is". These days hiker-bloggers may well become intimately familiar with the buzz words and prevalent narrative structures used by other hiker-bloggers prior to starting their own hike/blog, and thus with a good understanding of what they are expected to produce, and how they are expected to relate to certain species on the trail.

Storytelling and environment

It is clear from blogger narratives that there was a third main character in the master narrative of all hiker-bloggers, after themselves and the bear, snake or moose that they had encountered, which was the wild environment of the AT itself. Schiffrin (2009: 423) describes narrative as providing opportunities for constructing the self/other identity on two planes, one through the storytelling performance and one through the storyworld itself. She continues that, when space is personified as a protagonist or as crucial to the plot, "narrative may be more about place than about the people who populate it" (2009: 423). Sugiyama (2005: 186) describes narrative events as rooted in space, which simultaneously "constrains and enables" the actions that take place there. Interestingly, the example that she uses is of Little Red Riding Hood, in which the setting plays a large part in establishing Red Riding Hood's vulnerability: "she is travelling through woods where there are likely to be dangerous wild animals and where there are few if any people nearby to help her" (2005: 186). In writing about Londoners who blog about life in the city, Reed (2008: 398) emphasises the vital role that London plays in the events narrated, saying that, "for bloggers, each one of these events becomes another example of the city acting upon them and thus making them feel its presence". Similarly, the AT was not a 'backdrop' to interactions between hikers and wildlife; it played a distinct part in how people perceived that wildlife. Indeed, some hikers admitted that they looked at certain species differently on the trail to the way they viewed the same species at home.

Blogs as research data

The idea of researching a community without ever actually entering it is nothing new, and was written about extensively by Mead and Métraux, amongst others, in *The Study of Culture at a Distance* (2000 [1953]), a compilation of essays describing the techniques that US-based anthropologists used to study foreign cultures during the Second World War and since. At that time researchers used literature, film, art, informant interviews, games, slang, and any other sources that they could get their hands on in order to gain insights into the culture they were interested in. Mead (2000 [1953]: 11) wrote that the job of the researcher was to go beyond the source materials, in order to be able to "delineate in terms of a larger whole, the culture...". More recently, the possibilities for studying culture at a distance have become almost limitless, with the advent of the "global autobiography project" that is the internet (Murray, 1997: 252). In the relatively brief amount of time during which the Web has become integrated into people's daily lives, much has been written about it (see Azariah, 2016; Boellstorff et al., 2012; Garcia et al., 2009; Hine, 2000, 2009, 2015; Law and Yang, 2007; Lee, 2000; McNeill, 2003; Nakamura, 2002; Orgad, 2009; Postill and Pink, 2012; Reed, 2005, 2008; Schmidt, 2007; Trainer et al., 2016; Turkle, 1995; Zylinska, 2012), with a particular recent emphasis on social media (Armstrong and Koteyko, 2011; Blank and Reisdorf, 2012; Gross and Acquisti, 2005). Lee (2000: 14) has described how the growth of the internet has "encouraged renewed interest in how researchers 'forage' for sources of documentary data", suggesting that it is possible to treat online repositories as "field sites" and to adapt typical sampling strategies of field work to the documentary work done on these new "field sites". In drawing attention to an implicit bias against this type of work, Lee cites Catherine Hakim, who "makes the point that methodological discussion in social science research is oriented to the collection of new data as opposed to the use of existing materials" (2000: 85). Lee argues, however, that "data obtained opportunistically should not be seen as inherently inferior to data designed for a particular purpose" (2000: 9).

Bloggers tend to view the internet, or more specifically the blogosphere that they frequent on the internet, as a community to which they belong. Hine touches on this in arguing that internet use is embedded in everyday life, as opposed to existing in a space apart from it:

> We do not necessarily think of 'going online' as a discrete form of experience, but we instead often experience being online as an extension of other embodied ways of being and acting in the world...experience of the online environment has become seamlessly integrated with other embodied experiences...
>
> (Hine, 2015: 41)

Hine continues by describing how the internet has become a place for expressing an embodied self, rather than for leaving the body behind. Thus, rather than going online to 'reconstruct' their experiences with animals in the world of the AT, bloggers viewed being online and blogging about the AT as *a part* of their experience of the AT, in a similar way to how Reed's blogging Londoners (2008) viewed their blogging practices as part of their experience of dwelling in the city. Accordingly, when posting a blog on the Trek, hikers are not just posting *about* their actions in a community, they are directly acting within that community. Yet Reed (2005: 227) argues that the notion of online activity as taking place in a world which exists *alongside* the offline one (as opposed to as a part of the offline one) can lead bloggers to viewing their blogging activities as a means of "doubling the subject, having the 'I' of the prototype appear twice, occupying two separate space-times". This separate-but-connected notion of the internet in relation to the offline world transforms bloggers' acts of self-telling into acts of self-making, enabling them not only to 'double' themselves as subjects, but to remake themselves in the process. As Hine points out in talking about virtual ethnography, which she describes as "necessarily partial", "a holistic description of any informant, location or culture is impossible to achieve...Our accounts can be based on ideas of strategic relevance rather than faithful representations of objective realities" (2000: 65).

A note on hiker blogging practices

The circumstances of long-distance trail hiking do not lend themselves naturally to the habit of regular blogging, and it is therefore worth briefly describing the particular blogging practices performed by AT thru-hikers. People carry a variety of technologies with them on the trail. The vast majority of hikers will have a smartphone, and many people also take an iPad or similar tablet. One hiker, named Gadget, had been given the trail name due to the fact that he had taken a smartphone, iPad, camera, and laptop onto the trail with him;

considered by most to be an excessive amount of tech. Nevertheless, a hiker who blogs on their smartphone might also take a portable pocket keyboard with them, and many people spoke of carrying portable chargers, which mitigated the need to charge in town too regularly, as well as USB cables and other accessories. Some bloggers, writing for other hikers, recommended putting phones on airplane mode to save battery in between town visits. Wi-fi is generally not available in the backcountry. For this reason, many bloggers talked about composing their posts at night along the trail, and then uploading one or multiple posts whenever they were next able to access wi-fi, which usually entailed going into a nearby town, where service could be accessed at a hostel, inn, restaurant, camp store or library. Good wi-fi service is particularly important for longer posts (there is no word limit set on blog posts), or when someone wants to attach photographs to their blog post, which the vast majority of bloggers did. These photos would almost always also have been taken on their smartphone, sometimes with the help of a selfie stick attachment specifically designed to fit to a trekking pole (people liked to include photos of themselves in front of spectacular vistas, or sharing the frame with a wild animal or group of new hiker friends).

In terms of data collection, a blogosphere is a fluid entity, and the Trek blogosphere is no different. People add blogs every day, edit posts to say something slightly different, and occasionally even remove entire posts from their archives, once in a while extracting themselves – and all of their posts – from the blogosphere completely, so that it appears that they were never there. Meanwhile, other bloggers still post narratives about their AT adventure even years after completing it. Some bloggers change the name that they are writing under, switching from their real name to a 'hiker name', or from one old hiker name to a newly preferred handle, making it appear that one person has disappeared, or that two separate people have been blogging independently of each other. This means that pinning down specific numbers, for example, how many hiker-bloggers are online per year, or how many posts each wrote, can be challenging. Further, the content of each blog is wholly up to the person writing it, and so one person may have several very interesting (for the researcher!) encounters with wild animals but never write about them because they're far more interested in the other humans on the trail or what restaurants they stopped at, while another person may have one experience so meaningful to them that they proceed to write several blogs on the topic. Therefore, this book attempts to balance the need for an idea of numbers – for example, what proportion of hikers wrote about being charged by a bear – with the practicalities of an ever-morphing web of narratives.

Conclusion

The AT is a long and diversely populated woodland corridor which thousands of people hike along every year, living for several months in the woods, and alongside their other inhabitants. The trail is a unique site for interactions between wild animals and humans unaccustomed to the presence of autonomous animals, and the way in which hikers encounter and interact with these nonhuman persons is deeply interesting from an anthrozoological perspective. The thriving community of hikers who blog about their experiences on the trail allows for an exploration of these kinds of encounters. Although the online collection of blogged narratives is ever-expanding, this book delves into 1,691 of these narratives in order to find themes common to the experiences of thru-hikers with the other animals on the trail. The following chapters constitute an exploration of these themes.

Chapter 1 looks at how hiker perceptions of the 'meaning' of their hike influenced their perceptions of the animals that they met, by considering how hikers who viewed themselves as being on a pilgrimage perceived and talked about the nonhumans that they encountered along the way as embodying important elements of a nature pilgrimage. Chapter 2 is devoted to the black bears (*Ursus americanus*) who live on and around the trail, and whose interactions with AT hikers are increasingly fraught as they become accustomed to being able to obtain hiker food. Despite the reported incidents of conflict between bears and hikers, the presence of bears on the trail is highly valuable to hikers, not least because of their status as symbols of 'authentic' wilderness. In Chapter 3, the surprising reactions of hikers to many of the species on the trail as 'cute' is explored. As with so many of the themes arising from hiker blogs, there is a tension between cultural learnings about how to respond to animals, and in-the-moment embodied reactions to the Otherness of animals on the trail. Having spent time on the reassuring perception of trail animals as being cute, Chapter 4 turns to unpleasant and uncomfortable encounters with animals on the trail. The context in which certain species inspired feelings of disgust or fear in hikers is discussed, as well as looking at how encounters with humans may have proven unpleasant or uncomfortable for the nonhuman animal. Finally, in Chapter 5, hiker narratives describe the soundscape of the trail, including the multiplicity of animal sounds. Travelling through dense woodland, hikers frequently heard animals without seeing them, and so came to know the animals of the trail through the sounds that they made.

References

Abram, D. 1997. *The Spell of the Sensuous*. New York: Vintage Books.

Almagor, U. 1990. Odor and Private Language: Observations on the Phenomenology of Scent. *Human Studies* 13(3): 253–274.

Anderson, K. 1998. Animals, Science, and Spectacle in the City. In: J. Wolch and J. Emel (eds.) *Animal Geographies: Place, Politics and Identity and the Nature–Culture Borderlands*, pp. 119–138. London: Verso.

Archer, J. and Monton, S. 2010. Preferences for Infant Facial Features in Pet Dogs and Cats. *Ethology* 117: 217–226.

Arluke, A. and Sanders, C. R. 1996. *Regarding Animals*. Philadelphia: Temple University Press.

Armstrong, N. and Koteyko, N. 2011. "Oh dear, should I really be saying that on here?": Issues of Identity and Authority in an Online Diabetes Community. *Health* 16(4): 347–365.

ATC (Appalachian Trail Conservancy). 2014. *Appalachian Trail Conservancy Strategic Plan.*

ATC (Appalachian Trail Conservancy). 2016. Available at: www.appalachiantrail.org

Azariah, D. R. 2016. The Traveller as Author: Examining Self-Presentation and Discourse in the (Self) Published Travel Blog. *Media, Culture & Society* 38(6): 934–945.

Badone, E. and Roseman, S. R. 2004. Approaches to the Anthropology of Pilgrimage and Tourism. In: E. Badone and S. R. Roseman (eds.) *Intersecting Journeys: The Anthropology of Pilgrimage and Tourism*, pp. 1–23. Chicago: University of Illinois Press.

Baker, S. 2001. *Picturing the Beast.* Chicago: University of Illinois Press.

Baptiste, M. E., Whelan, J. B. and Frary, R. B. 1979. Visitor Perception of Black Bear Problems at Shenandoah National Park. *Wildlife Society Bulletin* 7(1): 25–29.

Berger, J. 1980. *About Looking.* London: Bloomsbury.

Berger, P. and Luckmann, T. 1991. *The Social Construction of Reality.* London: Penguin Books.

Bernard, H. R. 1995. *Research Methods in Anthropology.* Walnut Creek: AltaMira Press.

Besnier, N. 2016. Humour and Humility: Narratives of Modernity on Nukulaelae Atoll. *Etnofoor* 28(1): 75–95.

Bird-David, N. 1999. "Animism" Revisited: Personhood, Environment and Relational Epistemology. *Current Anthropology* 40(Supplement, February): 67–91.

Black, J. B. and Bower, G. H. 1980. Story Understanding as Problem Solving. *Poetics* 9: 223–250.

Blank, G. and Reisdorf, B. C. 2012. The Participatory Web. *Information, Communication & Society* 15(4): 537–554.

Boellstorff, T., Nardi, B., Pearce, C. and Taylor, T. L. 2012. *Ethnography and Virtual Worlds: A Handbook of Method.* Princeton: Princeton University Press.

Boulware, D. R. 2004. Gender Differences among Long-Distance Backpackers: A Prospective Study of Women Appalachian Trail Backpackers. *Wilderness and Environmental Medicine* 15(3): 175–180.

Bourdieu, P. 1986. The Forms of Capital. In: J. E. Richardson (ed.) *Handbook of Theory of Research for the Sociology of Education*, pp. 241–258. New York: Greenword Press.

Bratton, S. P. 2012. *The Spirit of the Appalachian Trail: Community, Environment, and Belief on a Long-Distance Hiking Path.* Knoxville: The University of Tennessee Press.

Brubaker, S. J. and Wright, C. 2006. Identity Transformation and Family Caregiving: Narratives of African American Teen Mothers. *Journal of Marriage and Family* 68(5): 1214–1228.

Bruner, J. 2004. Life as Narrative. *Social Research* 71(3): 691–710.

Bryson, B. 1998. *A Walk in the Woods.* London: Transworld Publishers.

Bulbeck, C. 2005. *Facing the Wild.* London: Earthscan.

Buller, H. 2012. "One Slash of Light, Then Gone": Animals as Movement. *Editions de L'Ehess* 189: 139–153.

Bulliet, R. W. 2005. *Hunters, Herders, and Hamburgers.* New York: Columbia University Press.

Burghardt, G. M., Hietala, R. O. and Pelton, M. R. 1972. Knowledge and Attitudes Concerning Black Bears by Users of the Great Smoky Mountains National Park. *Bears: Their Biology and Management* 2: 255–273.

Candea, M. 2010. "I Fell in Love with Carlos the Meerkat": Engagement and Detachment in Human–Animal Relations. *American Ethnologist* 37(2): 241–258.

Caplan, P. 2010. Death on the Farm: Culling Badgers in North Pembrokeshire. *Anthropology Today* 26(2): 14–18.

Cater, E. 2006. Ecotourism as a Western Construct. *Journal of Ecotourism* 5(1): 23–39.

Clark, J. E., van Manen, F. T. and Pelton, M. R. 2003. Survival of Nuisance American Black Bears Released On-Site in Great Smoky Mountains National Park. *Ursus* 14(2): 210–214.

Cornelius, M. 1991. *Katahdin With Love*. Tennessee: Milton.

Curtin, S. 2005. Nature, Wild Animals and Tourism: An Experiential View. *Journal of Ecotourism* 4(1): 1–15.

Dale, J. P., Goggin, J., Leyda, J., McIntyre, A. P. and Negra, D. (eds.). 2017. *The Aesthetics and Affects of Cuteness*. New York: Routledge.

Davis, Z. 2012. *Appalachian Trials*. UK: Good Badger.

Dearden, P. 1990. Non-Consumptive Wildlife-Oriented Recreation: A Conceptual Framework. *Biological Conservation* 53: 213–231.

Despret, V. 2011. Experimenting with Politics and Happiness – through Sheep, Cows and Pigs. *Unruly Creatures: The Art and Politics of the Animal*. London Graduate School, Kingston University and the Centre for Arts and Humanities Research, Natural History Museum, London (14th June 2011). Audio available at: http://backdoorbroadcasting.net/2011/06/vinciane-despret-experimenting-with-politics-and-happiness-through-sheep-cows-and-pigs/

van Dooren, T. 2014. Care. *Environmental Humanities* 5: 291–294.

Dorwart, C. E., Moore, R. L. and Yu-Fai Leung. 2009. Visitors' Perceptions of a Trail Environment and Effects on Experiences: A Model for Nature-Based Recreation Experiences. *Leisure Sciences* 32(1): 33–54.

Duffus, D. A. and Dearden, P. 1990. Non-Consumptive Wildlife-Oriented Recreation: A Conceptual Framework. *Biological Conservation* 53: 213–231.

Erb, P. L., McShea, W. J. and Guralnick, R. P. 2012. Anthropogenic Influences on Macro-Level Mammal Occupancy in the Appalachian Trail Corridor. *PLoS ONE* 7(8). Available at: https://doi.org/10.1371/journal.pone.0042574

Evernden, N. 1992. *The Social Creation of Nature*. London: The Johns Hopkins University Press.

Fondren, K. M. 2016. *Walking on the Wild Side*. New Jersey: Rutgers University Press.

Ford, K., Nadolny, R., Stromdahl, E. and Hickling, G. 2015. Tick Surveillance and Disease Prevention on the Appalachian Trail. *Park Science* 32(1): 36–41.

Foucault, M. and Miskowiec, J. 1986. Of Other Spaces. *Diacritics* 16(1): 22–27.

Franklin, A. 1999. *Animals and Modern Cultures*. London: SAGE.

Freidt, B., Hill, E., Gomez, E. and Goldenberg, M. 2010. A Benefits-Based Study of Appalachian Trail Users: Validation and Application of the Benefits of Hiking Scale. *Physical Health Education Nexus (PHENex)* 2(1): 1–22.

Fullagar, S. 2000. Desiring Nature: Identity and Becoming in Narratives of Travel. *Cultural Values* 4(1): 58–76.

Gall Myrick, J. G. 2015. Emotion Regulation, Procrastination, and Watching Cat Videos Online: Who Watches Internet Cats, Why, and to What Effect? *Computers in Human Behavior* 52: 168–176.

Garcia, A. C., Alecea, I., Standlee, J. B. and Cui, Y. 2009. Ethnographic Approaches to the Internet and Computer-Mediated Communication. *Journal of Contemporary Ethnography* 38: 52–84.

Gerard, K., Graefe, A., Manning, R. and Bacon, J. 2003. An Examination of the Relationship between Leisure Activity and Place Attachment among Hikers along the Appalachian Trail. *Journal of Leisure Research; Urbana* 35(3): 249–273.

Golbeck, J. 2011. The More People I Meet, the More I Like My Dog: A Study of Pet-Oriented Social Networks on the Web. *First Monday* 16(2). Available at: http://firstmonday.org/ojs/index.php/fm/article/view/2859/2765

Goldenberg, M., Hill, E. and Freidt, B. 2008. Why Individuals Hike the Appalachian Trail: A Qualitative Approach to Benefits. *Journal of Experiential Education* 30(3): 277–281.

Gomez, E., Freidt, B., Hill, E., Goldenberg, M. and Hill, L. 2010. Appalachian Trail Hiking Motivations and Means-End Theory: Theory, Management and Practice. *Journal of Outdoor Recreation, Education and Leadership* 2(3). Available at: https://doi.org/10.7768/1948-5123.1043

Gore, M. L., Knuth, B. A., Curtis, P. D. and Shanahan, J. E. 2006. Stakeholder Perceptions of Risk Associated with Human–Black Bear Conflicts in New York's Adirondack Campgrounds: Implications for Theory and Practice. *Wildlife Society Bulletin* 34(1): 36–43.

Gross, R. and Acquisti, A. 2005. Information Revelation and Privacy in Online Social Networks (The Facebook Case). *Proceedings of the 2005 ACM Workshop on Privacy in the Electronic Society (WPES)*: 71–80.

Hall, A. 2000. *A Journey North.* Massachusetts: Appalachian Mountain Club Books.

Hill, E. 2014. Appalachian and Pacific Crest Trail Hikers: A Comparison of Benefits and Motivations. *Journal of Unconventional Parks, Tourism & Recreation Research* 5(1). Available at: https://digitalcommons.odu.edu/hms_fac_pubs/87/

Hill, E., Goldenberg, M. and Freidt, B. 2009. Benefits of Hiking: A Means-End Approach on the Appalachian Trail. *Journal of Unconventional Parks, Tourism & Recreation Research* 2(1): 19–27.

Hill, J., Curtin, S. and Gough, G. 2014. Understanding Tourist Encounters with Nature: A Thematic Framework. *Tourism Geographies* 16(1): 68–87.

Hine, C. 2000. *Virtual Ethnography.* London: SAGE.

Hine, C. 2009. Question One: How Can Qualitative Internet Researchers Define the Boundaries of Their Projects? In: A. Markham and N. Baym (eds.) *Internet Inquiry: Conversations About Method*, pp. 1–20. Los Angeles: Sage.

Hine, C. 2015. *Ethnography for the Internet.* London: Bloomsbury.

Hurn, S. 2010. What's in a Name? *Anthropology Today* 26(3): 27–28.

Hurn, S. 2012. *Humans and Other Animals.* London: Pluto Press.

Ingold, T. 1994. Preface to the Paperback Edition. In: T. Ingold (ed.) *What Is an Animal?*, pp. xix–xxiv. London: Routledge.

Ingold, T. 2011. *Being Alive.* London: Routledge.

Isbell, L. 2006. Snakes as Agents of Evolutionary Change in Primate Brains. *Journal of Human Evolution* 51: 1–35.

Isenberg, A. C. 2002. The Moral Ecology of Wildlife. In: N. Rothfels (ed.) *Representing Animals*, pp. 48–64. Bloomington: Indiana University Press.

Kates, J. 2013. A Path Made of Words: The Journalistic Construction of the Appalachian Trail. *American Journalism* 30(1): 112–134.

Keul, A. 2013. Embodied Encounters between Humans and Gators. *Social & Cultural Geography* 14(8): 930–953.

Klein, V. A. 2015. The nature of nature: Space, place and identity on the Appalachian Trail. Unpublished PhD thesis, Kent State University College, Ohio, US.

Kubo, T. and Shoji, Y. 2016. Demand for Bear Viewing Hikes: Implications of Balancing Visitor Satisfaction with Safety in Protected Areas. *Journal of Outdoor Recreation and Tourism* 16: 44–49. Available at: http://dx.doi.org/10.1016/j.jort.2016.09.004

Kyle, G., Graefe, A., Manning, R. and Bacon, J. 2010. Predictors of Behavioural Loyalty Among Hikers Along the Appalachian Trail. *Leisure Sciences* 26(1): 99–118.

Laforteza, E. M. 2014. Cute-ifying Disability: Lil Bub, the Celebrity Cat. *M/C Journal* 17(2). Available at: http://journal.media-culture.org.au/index.php/mcjournal/article/view/784

Law, P. P and Yang, K. 2007. The Blog and Civil Society in China. *Philippine Sociological Review* 55: 116–125.

Lee, R. M. 2000. *Unobtrusive Methods in Social Research*. Buckingham: Open University Press.

Leopold, A. L. 1968 [1949]. *A Sand County Almanac*. London: Oxford University Press.

Littlefield, J. and Siudzinsk, R. A. 2012. "Hike Your Own Hike": Equipment and Serious Leisure along the Appalachian Trail. *Leisure Studies* 31(4): 465–486.

LoBue, V. and DeLoache, J. S. 2008. Detecting the Snake in the Grass: Attention to Fear-Relevant Stimuli by Adults and Young Children. *Psychological Science* 19(3): 284–289.

Lorenz, K. 1971. *Studies in Animal and Human Behaviour, Volume 2*. London: Methuen.

Lyng, S. 1990. Edgework: A Social Psychological Analysis of Voluntary Risk Taking. *The American Journal of Sociology* 95(4): 851–886.

MacKaye, B. 1921. An Appalachian Trail: A Project in Regional Planning. *Journal of the American Institute of Architects* 9: 325–330.

MacLennan, J. and Moore, R. L. 2011. Conflicts between Recreation Subworlds: The Case of Appalachian Trail Long-Distance Hikers. *LARNeT – The Cyber Journal of Applied Leisure and Recreation Research* 13(1): ref 30.

Malamud, R. 1998. *Reading Zoos: Representations of Animals and Captivity*. New York: New York University Press.

McArthur Jope, K. 1983. Habituation of Grizzly Bears to People: A Hypothesis. *International Conference on Bear Research And Management* 5: 322–327.

McNamara, K. E. and Prideaux, B. 2011. Experiencing "Natural" Heritage. *Current Issues in Tourism* 14(1): 47–55.

McNeill, L. 2003. Teaching an Old Genre New Tricks: The Diary on the Internet. *Biography* 26(1): 24–47.

Méad, M. and Metraux, R. (eds.) 2000 [1953]. *The Study of Culture at a Distance*. New York: Berghahn Books.

Merleau-Ponty, M. 2014 [1945]. *Phenomenology of Perception*. Oxon: Routledge.

Milton, K. 2002. *Loving Nature*. London: Routledge.

Milton, K. 2005. Anthropomorphism or Egomorphism? The Perception of Non-Human Persons by Human Ones. In: J. Knight (ed.) *Animals in Person: Cultural Perspectives on Human–Animal Intimacies*, pp. 255–271. Oxford: Berg.

Mittlefehldt, S. 2010. The People's Path: Conflict and Cooperation in the Acquisition of the Appalachian Trail. *Environmental History* 15(4): 643–669.

Mittlefehldt, S. 2013. *Tangled Roots: The Appalachian Trail and American Environmental Politics*. Seattle: University of Washington Press.

Mulkens, S. A. N., de Jong, P. J. and Merckelbach, H. 1996. Disgust and Spider Phobia. *Journal of Abnormal Psychology* 105(3): 464–468.

Murray, J. 1997. *Hamlet on the Holodeck: The Future of Narrative in Cyberspace.* New York: Simon & Schuster.

Nakamura, L. 2002. *Cybertypes: Race, Ethnicity and Identity on the Internet.* London: Routledge.

Nash, R. 1982. *Wilderness and The American Mind.* New Haven: Yale University Press.

Newman, P., Manning, R., Bacon, J., Graefe, A. and Kyle, G. 2003. An Evaluation of Appalachian Trail Hikers' Knowledge of Minimum Impact Skills and Practices. *International Journal of Wilderness* 9(2): 34–38.

Ngai, S. 2012. *Our Aesthetic Categories: Zany, Cute, Interesting.* Cambridge: Harvard University Press.

Nisbett, N. and Hinton, J. 2005. On and Off the Trail: Experiences of Individuals with Specialized Needs on the Appalachian Trail. *Tourism Review International* 8(3): 221–237.

NPS (National Park Service). 2008. *Appalachian National Scenic Trail Resource Management Plan 2008.* Available at: www.nps.gov/appa/learn/management/upload/Appalachian_Trail_Resource_Management_Plan.pdf

Oelschlaeger, M. 1991. *The Idea of Wilderness.* New Haven: Yale University Press.

Öhman, A. and Mineka, S. 2003. The Malicious Serpent: Snakes as a Prototypical Stimulus for an Evolved Module of Fear. *Current Directions in Psychological Science* 12(1): 5–9.

Orams, M. B. 2000. Tourists Getting Close to Whales, Is It What Whale-Watching Is All About? *Tourism Management* 21: 561–569.

Orams, M. B. 2002. Feeding Wildlife as a Tourism Attraction: A Review of Issues and Impacts. *Tourism Management* 23: 281–293.

Orgad, S. 2009. How Can Researchers Make Sense of Issues Involved in Collecting and Interpreting Online and Offline Data? In: A. Markham and N. Baym (eds.) *Internet Inquiry: Conversations About Method*, pp. 33–53. London: Sage.

Pelton, M. R. 1972. Use of Foot Trail Travellers in the Great Smoky Mountains National Park to Estimate Black Bear (Ursus americanus) Activity. *Bears: Their Biology and Management* 2(23): 36–42.

Pelton, M. R., Scott, C. D. and Burghardt, G. M. 1976. Attitudes and Opinions of Persons Experiencing Property Damage and/or Injury by Black Bears in the Great Smoky Mountains National Park. *Bears: Their Biology and Management* 3: 157–167.

Postill, J. and Pink, S. 2012. Social Media Ethnography: The Digital Researcher in a Messy Web. *Media International Australia* 145(1): 123–134.

Reed, A. 2005. "My Blog Is Me": Texts and Persons in UK Online Journal Culture (and Anthropology). *Ethnos* 70(2): 220–242.

Reed, A. 2008. "Blog This": Surfing the Metropolis and the Method of London. *The Journal of the Royal Anthropological Institute* 14(2): 391–406.

Reed, B. C. and Rasnake, M. S. 2016. An Assessment of Coliform Bacteria in Water Sources Near Appalachian Trail Shelters Within the Great Smoky Mountains National Park. *Wilderness and Environmental Medicine* 27(1): 107–110.

Reynolds, P. C. and Braithwaite, D. 2001. Towards a Conceptual Framework for Wildlife Tourism. *Tourism Management* 22: 31–42.

Rothfels, N (ed.). 2002. *Representing Animals.* Bloomington: Indiana University Press.

Russell, C. L. 1995. The Social Construction of Orangutans: An Ecotourist Experience. *Society and Animals* 3(2): 151–170.

Salleh, A. 1996. The Politics of Wilderness: Aborigines and Eco-Activists. *Arena Magazine* 23: 26–30.

Schänzel, H. A. and McIntosh, A. J. 2000. An Insight into the Personal and Emotive Context of Wildlife Viewing at the Penguin Place, Otago Peninsula, New Zealand. *Journal of Sustainable Tourism* 8(1): 36–52.

Schiffrin, D. 2009. Crossing Boundaries: The Nexus of Time, Space, Person, and Place in Narrative. *Language in Society* 38(4): 421–445.

Schmidt, J. 2007. Blogging Practices: An Analytical Framework. *Journal of Computer-Mediated Communication* 12: 1409–1427.

Shriver, G., Maniero, T., Schwarzkopf, K., Lambert, D., Dieffenbach, F., Owen, D., Wang, Y. Q., Nugranad-Marzilli, J., Tierney, G., Reese, C. and Moore, T. 2005. *Appalachian Trail Vital Signs. Technical Report MPS/NER/NRTR—2005/026.* Boston: National Park Service. Available at: http://npshistory.com/publications/appa/nrtr-2005-026.pdf

Singer, F. J. and Bratton, S. P. 1980. Black Bear/Human Conflicts in the Great Smoky Mountains National Park. *Bears: Their Biology and Management* 4: 137–139.

Smith, J. W., Moore, R. L. and Sommerville, M. 2013. Non-Sovereign Visitor Satisfaction: A Case Study of Military Training on the Appalachian Trail. *Managing Leisure* 18(3): 239–251.

Sugiyama, M. S. 2005. Reverse-Engineering Narrative: Evidence of Special Design. In: J. Gottschall and D. S. Wilson (eds.) *The Literary Animal: Evolution and the Nature of Narrative*, pp. 177–198. Evanston: Northwestern University Press.

Tate, J. and Pelton, M. R. 1983. Human–Bear Interactions in Great Smoky Mountains National Park. *Bears: Their Biology and Management*, Vol. 5, A Selection of Papers from the Fifth International Conference on Bear Research and Management, Madison, Wisconsin, USA.

Terry, D. P. and Vartabedian, S. 2013. Alone but Together: Eminent Performance on the Appalachian Trail. *Text and Performance Quarterly* 33(4): 344–360.

Thoreau, H. D. 1995 [1854]. *Walden; Or, Life in the Woods*. Mineola: Dover.

Trainer, S., Brewis, A., Wutich, A., Kurtz, L. and Niesluchowski, M. 2016. The Fat Self in Virtual Communities: Success and Failure in Weight-Loss Blogging. *Current Anthropology* 57(4): 523–528.

The Trek. 2016. Available at: www.thetrek.co/the-bloggers/?y=2016

The Trek. 2017. Available at: www.thetrek.co/the-bloggers/?y=2017

Tuan, Y. 1977. *Space and Place*. Minneapolis: University of Minnesota Press.

Turkle, S. 1995. *Life on the Screen*. New York: Touchstone.

Turner, J. 1996. *The Abstract Wild*. Tucson: The University of Arizona Press.

Ulrich, R. S. 1983. Aesthetic and Affective Response to Natural Environment. In: I. Altman and J. F. Wohlwill (eds.) *Behaviour and the Natural Environment*, pp. 85–126. New York: Plenum Press.

Urry, J. 1990. *The Tourist Gaze*. London: SAGE.

West, P. and Carrier, J. G. 2004. Ecotourism and Authenticity: Getting Away from It All? *Current Anthropology* 45(4): 483–498.

Wilson, E. O. 2005. Foreword from the Scientific Side. In: J. Gottschall and D. S. Wilson (eds.) *The Literary Animal: Evolution and the Nature of Narrative*, pp. vii–xii. Evanston: Northwestern University Press.

Wolch, J. and Emel, J. (eds.) 1998. *Animal Geographies: Place, Politics and Identity and the Nature–Culture Borderlands*. London: Verso.

Zarnoch, S. J., Bowker, J. M. and Cordell, K. 2011. A Mixed-Modes Approach for Estimating Hiking on Trails Through Diverse Forest Landscapes: The Case of the Appalachian Trail. *Canadian Journal of Forest Research* 41(12): 2346–2358.

Zylinksa, J. 2012. What If Foucault Had Had a Blog? In: K. B. Wurth (ed.) *Between Page and Screen: Remaking Literature Through Cinema and Cyberspace*, pp. 62–74. New York: Fordham University Press.

1 Hiking as pilgrimage, and wild animals on the trail

We need the tonic of wildness, – to wade sometimes in marshes where the bittern and the meadow hen lurk, and hear the booming of the snipe; to smell the whispering sedge where only some wilder and more solitary fowl builds her nest, and the mink crawls with its belly close to the ground...We need to witness our own limits transgressed, and some life pasturing freely where we never wander.

(Thoreau, 2007 [1861]: 205)

Don't be afraid of stepping away from your life if something in the wilderness compels you.

(Appalachian Trail hiker)

Why pilgrimage?

A hiker's perception of the nonhuman animals on the AT was informed in part through her thoughts about why she was on the trail and what she wanted to achieve there; thoughts which provided a framework for how encounters with animals were experienced and how the animal himself was talked about. One of the common ways of approaching a thru-hike of the AT was as a pilgrimage, and encounters with animals on the trail were regularly incorporated into the pilgrimage narratives of AT hikers. The perception and experience of being on a pilgrimage, be it an explicitly religious pilgrimage, a 'nature pilgrimage', 'wilderness pilgrimage' or more generally a pilgrimage of 'self-realisation' (see Bratton, 2012; Fondren, 2016), influenced how people engaged with animals on the trail, and how they remembered the experience afterwards in their narratives.

The Appalachian trail [is] a 2,150 mile long 'linear cathedral'.

(Terry and Vartabedian, 2013: 343)

Religious pilgrimage as a practice has been studied and written about extensively (see Badone and Roseman, 2004; Coleman and Elsner, 1995; Collins-Kreiner and Kliot, 2000; Frey, 2004; Holmes-Rodman, 2004; Lois Gonzalez, 2013; Nolan and Nolan, 1992; Rinschede, 1992; Roseman, 2004;

Sallnow and Eade, 1991; Timothy and Olsen, 2006; Turner and Turner, 1978; Vukonic, 1996). The notion of secular pilgrimage has also been very popular in recent years, particularly with scholars of tourism, who have drawn parallels between religious pilgrimage and apparently secular touristic practices such as ecotourism, mountain climbing, wilderness walking, dark tourism and even *Star Trek* convention attendance (see Collins-Kreiner, 2016; Goodnow and Bloom, 2017; Knox and Hannam, 2015; MacCannell, 1973; Porter 2004). Writing on ecotourism in particular has shown that it possesses many of the recognised elements of pilgrimage, including a quest for authenticity (see Bulbeck, 2005; Curtin, 2005; Schänzel and McIntosh, 2000; West and Carrier, 2004), spiritual rejuvenation or transformation (see Bulbeck, 2005) and feelings of awe and wonderment (see Fullagar, 2000; Schänzel and McIntosh, 2000). Yet Knox and Hannam put forward the possibility of arguing that engagement with *any* type of tourism can be viewed as an act of pilgrimage, "in which the 'faith' or 'doctrine' being reinforced is the notion of the holiday or trip as an escape, as a punctuation mark or as a *rite of passage*" (2015: 46, their emphasis). Their argument that even snowboarders are on a quest for 'authenticity' when on holiday would seem to take the comparison to traditional religious pilgrimage so far as to render comparisons almost meaningless.

In light of this, it is unsurprising that hiking the more than 2,000 mile long AT has been likened to a pilgrimage, particularly in Bratton's *The Spirit of the Appalachian Trail* (2012), which examines the role of religious and spiritual belief in the lives of hikers on the trail. Through interviews and an extensive paper survey, Bratton (2012: xvii) found that a variety of spiritual perspectives were expressed by long-distance hikers on the AT. Some people sought wilderness as an escape from society as a whole – including the perceived pressures of organised religion – others viewed the wilderness as a way to be closer to the God of their chosen denomination, while others engaged in a range of meditative or reflective practices on the trail that were not linked to any specific religion. Bratton also makes clear that not every hiker experienced a spiritual dimension to their hike; some saw their trek simply as an athletic challenge. In response to the paper survey that she distributed at hostels along the trail, 62 percent of people indicated that they were having spiritual or religious experiences, 32 percent said they were not and 6 percent did not answer the question (2012: 168).

It is clear from hiker blogs, too, that many people on the trail viewed their hike as either pilgrimage-like or as having pilgrimage-like attributes – and that this influenced how they met autonomous animals on the trail. It is worth noting that while only a minority of people writing on the Trek expressed this experience of pilgrimage in an overtly religious way, many more expressed their approach to pilgrimage on the AT as having a sacred or transcendent element to it, and, as with Bratton's findings, some did not refer to spirituality at all. Therefore, to talk about the pilgrimage experience of thru-hikers is to refer to something that was far from a homogenous experience. Yet parallels to traditionally religious pilgrimage included the pilgrim's/hiker's desire for a connection with something greater than themselves and their ensuing search for authenticity, their voluntary removal from everyday life, and the associated sacrifice of being away from any luxuries they were used to, the state of liminality entered into by leaving society behind and embarking on the journey, the physical act of walking many miles and the toll that hardship took on their body, and the sense of community that often formed around groups of pilgrims/hikers who shared similar worldviews and were all focused on the same goal. Again, not *all* thru-hikers who hit the trail experienced their journey as a pilgrimage, but many of them did. Furthermore, those who did view their hike as akin to a pilgrimage are unlikely to have been continuously engaged with this notion, meaning that they will have spent time 'pilgrimaging' and time 'not pilgrimaging'.

Inspiration

In "Pilgrimage and Its Aftermaths" (2004) Frey describes the motivations behind religious pilgrimage:

> Many pilgrims have difficulty articulating the reasons for their journey. Pilgrims are often attracted to the metaphorical pilgrimage as an inner journey and want to actualize the physical journey to help them access those inner "destinations" that are distant in daily life. Other pilgrims specifically plan and make the journey as a break from their daily lives; as a way to cope with the death of a loved one; as a transition from other serious losses, such as divorce, separation, or unemployment; to fulfil a religious vow; or to have a period of reflection and decision making. Often motives are multi-layered and evolve over the course of the journey.
>
> (Frey, 2004: 91)

From Frey's description it is clear that Christian pilgrimage (her focus is on the Camino de Santiago, which ends at the shrine of the apostle St James the Great in Spain) has complex reasons behind it, aside from the pilgrim's assumed religious devotion, that might equally apply to secular or spiritual (but not strictly religious) pilgrimage. Many of the reasons that she lists could be interpreted as a quest to find, or re-find, meaning in a person's life. Cohen, in his phenomenology of tourist experience, describes this quest as a search for the "centre" (1979: 180). He cites Eliade, who defined the 'centre' as "...pre-eminently the zone of the sacred, the zone of absolute *reality*" (Cohen, 1979: 180, my emphasis). Cohen conceives of different modes of touristic experiences, which reflect in part where the 'centre' is considered to be located. For many people the centre is at home, in their community, and when they go away they are simply looking for a recreational experience that will rejuvenate them before they return, refreshed, back to their centre of meaning. The pilgrim, on the other hand, locates the centre as 'out there' and undertakes their pilgrimage in order to find it. Importantly, there are 'modes of recreation' between these two extremes that describe a traveller who is neither completely a recreational tourist, nor a wholly focused pilgrim, but somewhere in between, and often depending on what stage of their journey they are on. This is a helpful way of interpreting the experience of AT thru-hikers, many of whom appear to have occupied different modes of recreation from day to day and even moment to moment, rarely being 'fully pilgrim' or 'fully not-pilgrim'.

Blog posts written by AT hikers or aspiring hikers very much reflected both Frey's and Cohen's notions about why people embark on a pilgrimage. The most 'mystical' of these narratives evoked the initial idea of the thru-hike as coming out of sudden and unexpected inspiration. One blogger talked about watching the sun set one day, when "I stared off into the hills and heard a voice say quite clearly, 'I want to go live in the woods for a while'". Another hiker, who referred to himself as "not an overly religious man", described the "epiphany" he had experienced on first deciding to hike the trail.

> Four years ago, sometime in the middle of winter, it came more like a whisper...not an apparition or a vision...more like a thought... "thru-hike the AT". Whattt? Say again? "Thru-hike the AT". Oh I get it...hike the Appalachian Trail. Where did that come from?

Another described being 'called' to walk to Katahdin, the mountain in Maine that forms the official northern summit of the trail.

> [N]othing in my life has ever felt so right. I feel Katahdin calling and I know I'll kick myself every day for the rest of my life if I don't try to accomplish this goal. I can't ignore this kind of feeling... It's well, it's more than a feeling. It's a mission.

Other bloggers talked about specific reasons for hiking, describing both what they wanted to leave behind them, and what they wanted to find out on the trail. The things that they were hiking to 'get away from' often evoked powerful feelings of negativity, and revolved around what we might consider the trappings of modern life; unfulfilling, routinised jobs, a focus on earning and consumption at the expense of quality of life, overwhelmingly pessimistic news stories, politics, the fast pace of life in general. Bratton (2012: 20) points out that the AT is a "protected space exempt from economic productivity". As one blogger put it, she sometimes felt overwhelmed by "strip malls and cars and rampant consumerism". For many people, the notion of a thru-hike appears to have been just as much about what they wanted to escape from as what they wanted to find.

> I hike because I'm tired of sitting around my house trying to figure out where my life is going. I'm tired of wasting my life away playing the life-sucking game Candy Crush while binge-watching Netflix and Hulu. I'm

tired of the routine. I'm tired of looking at the mass amount of material shit I have around my house that really adds nothing to my life. I'm tired of hearing and seeing how unhappy people really are.

Another blogger described the idea to thru-hike coming to her from a book she was reading.

[T]here was a short passage that alluded to the transformative powers of a thru-hike. It essentially stated that the sun, filtering through the trees on the Appalachian Trail, changes your body chemistry. I was at a point in my life where I was spending 50 hours a week inside the walls of a cubicle. I thought that if the sun could alter me, the artificial light in my cubicle must have been turning me into a piece of slime. If I remained shackled to my desk, I might have morphed into a corporate hunchback, permanently at the disposal of the 'You've Got Mail' bell. It was in this moment of clarity that I knew I had to witness the metamorphic powers of the trail for myself.

One hiker quoted a passage written by the wilderness advocate John Muir, responding, "I do feel like I am losing precious days. I do feel like I am just a machine for making money". Another said, "We are expected to work until there is no joy left in life. We are expected to put ourselves in debt until we die. Hiking the trail is my rebellion" and "one feeling I never get in the middle of the city, or in a town, or even out in the sticks where I live, I don't feel free". A blogger talked about how a day spent away from the internet had caused her severe anxiety, leading her to realise that she had to spend "some significant time in the wilderness" (although not commenting on her associated decision to blog about it). Others also mentioned undertaking the hike in order to tackle anxiety and depression.

These types of descriptions align with Fondren's findings in her recent ethnography of the AT (2016). Her informants talked about being miserable in their jobs, unhappy with the pressure to conform to structured schedules and routines, overwhelmed by the constant flow of information from various media, and exhausted by their consumption-driven lifestyles. The strong desire to cut ties with their old lives and start afresh can be seen in one Trek blogger's post.

Exactly two weeks ago [we] chopped our hair off and walked into the woods...I dropped our "baggage" alongside our hair in the woods. The birds will find better use for it than we will on the trail.

In *Escapism* (1998) Tuan describes the motivation behind 'extreme' actions like this as a reaction to human civilisation:

Life up there doesn't seem quite real...and although people do not mind living in a pleasant dream, they might well think there is something amiss

living and dying in one without ever knowing what it is to be awake...
Extremists yearn to sink to the bottom, to hug the earth, the gritty texture
and harshness that make for reality.

(Tuan, 1998: xiii)

Tuan's statement reinforces the types of binaries that bloggers evoked
throughout their narratives: society vs. nature, routine vs. freedom, consumption vs. minimalism, mindless vs. mindful, artificial vs. real, and as Tuan
suggests, asleep vs. awake. To a degree, bloggers seemed to take these notions
for granted, so that their thoughts about what they were getting 'away from'
set certain expectations about what they were going towards.

Looking for wild nature

Badone (2004: 184) describes how people commonly seek self-renewal
through "radical separation" from society and immersion in spaces belonging
to the Other. She writes that, "since Christianity no longer provides a com-
pelling means of redemption, alternative avenues are sought. These include
the encounter with alterity in the form of Nature..." (2004: 184). She continues: "these travellers seek a form of immortality or self-transcendence
through identification with a timeless heritage – cultural, architectural, and
material, or natural – that persists beyond the individual lifetime" (2004: 184).
Indeed, it was a connection with nature, and more specifically, wild nature,
that many bloggers talked about finding, and that could be considered for
many hikers as replacing the traditional institutions of religion that used to
be the guiding 'centre' of pilgrimage. For the minority of hikers who talked
about the thru-hike as a specifically religious (Christian) pilgrimage, nature in
this context was seen as the manifestation of God. As one wrote, "this appreciation extended to everything I encountered – other hikers, animals, even
Lichens! We are all living monuments to God".

For hiker-bloggers, the idea of living in wilderness was commensurate with
the idea of living a life of freedom, authenticity, and connection. Bloggers
wrote statements like: "Out there, it's freedom. Nobody judging you, just you,
the mountains, and the fellow thru-hikers who also feel free", and "through
nature we are all connected to one another". As the embodiment of a very specific set of values, wilderness was positioned in binary opposition to civilised
society, which was often described as being restrictive, inauthentic, and alien-
ating. As one blogger said, "Most of us visiting wilderness areas come from
urban areas small or large. We plan for months or years or decades to visit
these places – to 'get away'".

The history of wilderness

The way that hiker-bloggers wrote about wilderness is consistent with prevalent attitudes towards 'the wild' and 'nature' in the West, and particularly in

the US. Yet, far from being naturally occurring, these ideas have evolved over time and are historically and culturally determined. In *The Idea of Wilderness* (1991), Oelschlaeger finds that the notion of wildness originated around ten to 15 thousand years ago, as the Palaeolithic era phased into the Neolithic, and hunter-gatherers began to farm and herd. These new agriculturalists were the first people to make attempts at modifying their surroundings, and thus the concepts of 'nature' and 'culture' were conceived (1991: 28). As Ingold writes, "our ancestors crossed a threshold to culture, and in so doing launched themselves onto an entirely new plane of existence, ideational rather than physical" (1994: xix). Nature, no longer all that existed, could now begin to *mean something*.

For most of human history this meaning was as "an enemy to be conquered, a monster to be tamed" (Baron, 2004: 8). The first European immigrants to the New World "sought release from oppressive European laws and traditions, yet the complete license of the wilderness was an overdose" (Nash, 1982: 29). The new colonies of North America had to be constantly vigilant against the temptations towards wildness and the perceived barbarism that their new freedom and the surrounding wilds might animate. Thus, from the advent of agriculture up until the 19th century, wildness was thought of variously as wilful, terrifying, unruly, degenerate, subversive, a beast within us all, "purposeless" (Evernden, 1992: 30), and worthless. Wilderness was "dark and forbidding...ugly and crude" (McKibben, 2003: 50–52), "base and worthless" (Oelschlaeger, 1991: 68), "a chaotic sea...an earthly hell...[a] place in which a person feels stripped of guidance, lost and perplexed" (Nash, 1982: xii–3), the place of the original Fall, ugly, and a waste.

Yet when a census revealed that the American frontier had ended in 1890, "the scarcity theory of value began to work on behalf of wilderness" (Nash, 1982: xii). Wild lands were disappearing, and wildness itself seemed to be undergoing an extirpation. Gradually the wilderness took on worth where once it had been worthless, and people began to incorporate it into the realm of what they considered important in the world. Essays by wilderness advocates like Henry David Thoreau (1817–1862), John Muir (1838–1914), and Aldo Leopold (1887–1948) expounded the virtues of wild nature. For Thoreau, wildness represented "absolute freedom" (2007 [1861]: 7). He wrote that "life consists with wildness. The most alive is the wildest" (2007 [1861]: 28). Like Thoreau, Muir believed that humans needed wildness, and encouraged city dwellers to visit the wilderness for their own invigoration, saying that going into the wilderness was "going home" and that "the clearest way into the Universe...is through a forest wilderness" (Nash, 1982: 126). By the 20th century, 'wilderness' had acquired a multiplicity of relatively new meanings. This included, as Andrew C. Isenberg has argued, "no specific ecological meaning", instead representing "authenticity, freedom, virtue...godliness" (2002: 48).

In the last few decades the assumed binary of nature/culture has been repeatedly critiqued (see Cronon, 1996; Price, 1999; Raglon, 2009; Washington,

2007; Whatmore and Thorne, 1998; Wilson, 1992), yet ideas about wilderness as the locus of authenticity and freedom remain rooted in the popular consciousness. Salleh's description of wilderness highlights its ability to represent 'all things to all people':

> For the conformist, wilderness provides a chance to be delinquent; for the delinquent, a chance to conserve and protect. In contrast to an aggressive culture, wilderness is a dream of gentleness and peace; against materialism it is a spiritual transcendence. In a life-negating society, a river tells the phases of human life; in a sexually repressed society, wilderness recharges the senses, and "where emotion is denied, it speaks what is unfelt".
>
> (Salleh, 1996: 26)

Such socially ingrained ideas about wilderness were clearly an influence on people undertaking a thru-hike of the AT, whether or not 'being in nature' was their primary motivation for hiking (and it's important to remember that not everyone undertook a thru-hike with this priority). One blogger planning her thru-hike wrote, "Let's be honest, nature is the most healing power there is, it's 100 times more effective than therapy...Thinking about letting myself open up for nature to do its magic sounds like ultimate peace...". Another wrote, "We absorb the sanity in the wild order of Nature. We go there to be ordered by Nature". Another explicitly invoked the image of their thru-hike as pilgrimage, echoing Frey's description of pilgrimage as a physical journey intended to parallel an inner journey.

> Be transformed. This meant embarking on a pilgrimage on the AT as a metaphor for the journey of a family through life. As we overcame the hardships of the trail, we wanted to enjoy relationships over material possessions, find joy in simple pleasures, reflect in prayer, applaud the views, let the beauty of nature sink in, and be transformed through the laughter and the tears.

Authenticity

One of the hallmarks of pilgrimage is the search for authenticity. One blogger talked about people referring to her leaving behind the "real" world, arguing that "there's nothing more real than a journey through nature". Debates about the location of the 'real' world were a recurring theme in blog posts, some hikers referring to not wanting to go back to the real world, clearly viewing the society they were escaping from as the locus of the real, with others arguing that the woods were real and everything they had come from was 'artificial'. Aside from the binary created by these kinds of arguments it is helpful to look at them through Cohen's description of the 'centre'. For some, the centre (the real) was still located in the community they had come from, even if they wanted to avoid the centre for as long as possible. For others, the centre (the real) was located in the wilderness, leaving modern society as inevitably 'fake' in contrast. Merleau-Ponty (2014 [1945]: 299) describes how, while our body and our perception ask us to take the landscape that we are in as the centre of our world, it is possible for us, if we are kept away from what we love (where we feel the most 'connected') to feel decentred and destabilised. For some, the 'real' setting of the wilderness was viewed as the centre, in that it was intended as a background against which to 'awaken' to their 'real' selves.

> The solitude of nature is a scary thing. It's such a different world than the one I normally live in…What will it be like to spend a day alone in the depths of nature? What will I discover about myself when there's no one around to distract me? I hope to get to know myself better, and to find some peace.

Walking

> [T]o move, to know, and to describe are not separate operations that follow one another in a series, but rather parallel facets of the same process – that of life itself. It is by moving that we know, and it is by moving, too, that we describe. It is absurd to ask, for example, whether ordinary walking is a way of moving, knowing or describing. It is all three at once.
>
> (Ingold, 2011: xii)

At a most basic level, pilgrimage is about walking. The reasons for undertaking a pilgrimage can be complex and elusive, even to the person embarking on it, but the physical experience of walking for weeks and months is something tangible, literally providing a foothold for the pilgrim, something to be certain of. Scriven (2014: 252) examines pilgrimage from the perspective of the mobilities paradigm in the social sciences, claiming that "pilgrimage is produced, performed and experienced in corporeal movements and embodied actions that shape individual and communal interactions with the environment". Pilgrimage can be interpreted as an embodied performance of journeying. As one person wrote, not in a blog but in a post to the Appalachian Long Distance Hiker's Association page on Facebook, "I put my symptoms on WebMD and it turns out I just need to be on a trail hiking" (ALDHA Facebook Group, 2015: np). Frey emphasises the importance of the physical over all else in her study of the Santiago pilgrims:

> One's religious practice or motive is secondary to the more important element of the pilgrimage: how the pilgrim physically makes the journey itself. That is, the *how* often takes primacy over the *why* of the journey among Santiago pilgrims.
>
> (Frey, 2004: 91)

Likewise, Holmes-Rodman (2004), in her study of women pilgrims to Chimayo, New Mexico, writes about how the shrine itself was never mentioned by the women, unless it was in terms of how many miles away it was. Instead, their memories were of and about the journey itself: the sacrifice, the blisters, the pain of walking. As Fife (2004: 156) says, "the key sign seems to involve an effort of movement (both literal and metaphorical) – one powerful enough to change an indistinguishable space into a sacred place through desires expressed in the movement itself". One thru-hiking blogger wrote, "This year I've fallen in love with the idea of movement and migration", and another emphasised the clarity of purpose associated with walking the trail, saying, "when in doubt just walk North".

Darby, writing about her experience with walking clubs in England, states that "walking is a socially constitutive force shaping personal identity and a sense of community in the face of social fragmentation" (2000: 4), and frames the walking group as a rite of secular pilgrimage, pointing to "the help that is generated through re-establishing a physical grounding in self" (2000: 227). One AT blogger, a military veteran who described herself as suffering from PTSD, wrote that the only time she had been able to find peace since returning from deployment was to go hiking with her service dog.

Edensor (2000: 82) writes specifically about the significance of walking in nature, describing the practice as designed to achieve a reflexive awareness of the body. He explores the "distinctive ways in which we express ourselves physically, simultaneously performing and transmitting meaning while sensually apprehending 'nature', and the role of the body in nature" (2000: 82). It is this sense of feedback between their bodies and nature that bloggers frequently described. Fondren writes about this experience in her ethnography of the AT:

> When I asked him if there was a commonality among long-distance hikers, he hesitated a bit. He attempted to answer the question by saying, "It's the outdoors", but then asked, "Has anyone been able to explain it to you?"
> (Fondren, 2016: 19)

The desire to go on a wilderness pilgrimage is undoubtedly informed by cultural ideas about the meaning of 'wilderness', yet weeks or months spent on the trail, immersed in the wilderness of popular mythology, endowed people with a point of view based on their actual physical experiences of the environment. Hikers on or after their hikes often described a somatic, embodied experience of wilderness, arguably only partly informed by entrenched cultural notions of the meaning of wilderness.

> [S]tripping my life down to pure necessities and embarking alone into the wilderness allowed me to see things through a different set of eyes.
> Each day as I walk I wonder what to focus my attention on. Do I feel my body as it moves...Do I marvel at the intricacies of my feet and how they absorb information from the earth...What about listening to the ever changing rhythm of my breath and acknowledging the exchange that

constantly occurs between my inhale and the exhale of the wilderness surrounding me?

Being in town felt a bit bizarre. The neon signs and cars passing by were dissatisfying and mesmerising. Fluorescent light is bullshit compared to the rays (or misty days) that are provided in the wilderness.

One blogger wrote simply, "Why did I have such a strong reaction to being out in the wilderness?" Frey (2004: 101) describes how the landscapes of a pilgrimage become inscribed with the stories and experiences of the pilgrim, and are remembered in the body, as a type of "corporeal memory". She quotes one of her informants as commenting that her experience of pilgrimage "was recorded on my body" (2004: 101). Another blogger described the end of a long day of hiking.

I was so happy that my feet felt better…I made my way to Standing Indian Shelter with tears streaming down my face…just because I was happy…as I listened to the Wailin' Jennys singing "Alleluia". I truly felt blessed and connected to the universe as I finished up my long day with joy rather than exhaustion.

Transformation on the trail

Many people expressed a desire to experience some kind of transformation while on the trail, often either in order to 'find' themselves, or to become a 'better' self; stronger, braver, more independent, more resilient, more free-spirited, wiser, and/or more connected to the authentic. One 68-year-old blogger mused about why she was on the trail, saying, "I certainly didn't need to 'find myself', and I certainly didn't have anything to prove", perhaps reflecting an awareness of being in a minority among her fellow travellers, with some rhapsodising about the transformative experiences they were having.

Moving through the woods each day gives me insurmountable joy, and I find deep comfort sleeping under the stars every night. My thoughts are beginning to deepen (which I've realised is a process that can't be rushed) and I find myself wrapped up for hours in ideas of spirituality, my relationship with nature, and who I am as a being on this earth.

One person wrote about having been on anxiety medication for eight years before embarking on her thru-hike, during which she found that "the trail takes your stress, your anxiety and your negativity and absorbs it into the canopy". She wrote that although she had only been on the trail for one week, it had been enough to "drastically alter" her life. Another wrote about his renewed connection with nature.

I can tell you with no shame, there have been times on the Appalachian Trail when I felt like a squirrel or a bear, or an ancient tree with a great hollow space in its heart, or a big rock that outlasted the dinosaurs. Today, on some level I still understand myself this way. My name is squirrel. My name is bear. My name is rock. I am the mountain and the mountain is me. That sounds crazy but go walk two thousand miles in the woods... For the rest of your life, your sense of 'I am' will be changed forever. You will identify with the natural world, as you were born to do before the doctor slapped you on the ass and began to civilize you...

For others, the anticipated transformation did not come as easily as expected, with one blogger writing about her desire to leave the trail early, but still searching for the "Divine Feminine" and saying that she had spent all day "trying to open myself up to the feminine in nature all around me".

Many perceived their ongoing transformation as due not only to being out in nature, but to being there with other people, and the sense of community felt by sharing a common goal, and a common appreciation for the wild surroundings. One blogger wrote, "Six months on the trail crumbled the resistance and defences I'd spent a lifetime creating", citing the warmth and comfort she had found both in nature and in the other hikers on the trail. Another wrote that her time on the trail had "renewed my faith in humanity while also reminding me of the untameable power of Mother Nature, a power that repeatedly kicked me to the earth below". Olwig (1993: 327) describes the process by which people travel together "by primitive means" as naturally causing people to form ties with each other that cross social boundaries. Badone and Roseman (2004: 2) align with Durkheim in arguing that social collectivities can give rise to notions of divinity and the sacred, generating "religious" emotions. Fondren found that many of her informants admitted to choosing the AT (rather than other long-distance hiking trails, such as the Pacific Crest or Continental Divide trails) specifically because it is known as a "social" trail (2016: 6). From day to day a hiker has the option of choosing to hike alone in solitude, or to hike with their fellow pilgrims. Friendships are made quickly, and by most accounts, easily, but this does not constitute an obligation to stay with each other for the remainder of the hike.

In addition to meeting fellow thru-hikers on the trail, hikers regularly met people who they referred to, and who referred to themselves, as 'trail angels', and who provided 'trail magic'. In some places on the trail hikers came across trail angels every day, and in other places they could go for a week without meeting one. Trail angels are people who generally live near to the trail, and who help hikers on their journey, either through offering them a ride to town, by bringing food and drink onto the trail specifically so that they can donate it to any thru-hikers that they meet, or even by offering them a shower and a place to stay for the night. Bloggers told stories of complete strangers – trail angels – cooking them pancakes by the side of the trail, having all-day barbeques with all the trimmings, offering plates of sandwiches and cookies, fresh fruit and drinks. Sometimes the angel was nowhere to be seen, and the thru-hiker stumbled across a cooler filled with snack bars, beer and fizzy drinks propped up against a tree on the trail, which they knew they were welcome to help themselves from.

> The phenomenon known as trail magic has been overwhelming…as we sat along the road a woman arrived and asked if we were thru hikers. She was visiting a 73 year old thru hiking friend and grabbed a dozen McDonald's cheeseburgers and a bunch of cold sodas knowing hikers would be at this parking lot…It was perfect timing. I was honestly down in the dumps with no plan to get back out yet. What a genuinely awesome feeling.

The phenomenon of people who spend their own money and time to donate to strangers in this way speaks to the idea of the thru-hike as pilgrimage, and of the hikers as pilgrims who should be supported on their arduous journey. For many bloggers, notions of transformation were associated with the idea of sacrifice, but the generosity of trail angels was perceived as helping to make these sacrifices bearable. Overall, the experience of *communitas* that has been described as a central aspect of pilgrimage was viewed as supportive to the individual's quest for transformation.

The return home

One person wrote that, "the AT changes you and it fixes you", and another that the trail "helped me love myself again". Yet post-hike depression was a much talked-about topic in hiker blogs, and appears to have afflicted many thru-hikers on their return home from the AT. Hikers reported feeling tearful, despondent, claustrophobic, and overwhelmed by the pace and/or the perceived banality of the life that they had returned to.

> Well, I'm sad to report that the blues are very real. Normal life has hit me like a stack of bricks…My first night home I cried hysterically, but I couldn't say why. I wanted to say I missed my home…being away from that life was breaking my heart.

Frey (2004: 99) describes this time as a period of culture shock and adjustment that can be very difficult and disorienting for the pilgrim. One blog post warned recently returned hikers not to make any big decisions for a while after returning, as they may be rash. Another described her fears, and a seeming uncertainty about the location of her 'centre' as Cohen might phrase it.

> I'm scared of forgetting so much. I'm scared of losing what the trail taught me. I'm terrified of re-emerging into society and thinking that maybe I do need all this 'stuff'. That maybe this AT hike was just a 'phase' and that the 'real' world will always be waiting.

Morinis (1992) describes the return home as a component of the pilgrimage: "While the sacred place is the source of power and salvation, it is at home once again that the effects of power are incorporated into life and what salvation is gained is confirmed...Has there been change? Will it last?"

The magic of trail animals

The particular mind-set in which hiking the trail was akin to a pilgrimage crucially influenced how hikers perceived the animals that shared their lifeworlds with them on the trail, in what Fullagar describes as "encounters with nature's otherness which are structured by the paradoxical desire to experience difference whilst also recognising a common materiality" (2000: 59).

Pilgrims undertake their journey looking for a connection with the 'real', something bigger than themselves. In hiker narratives, 'nature' was frequently viewed as the real that hikers wanted to feel connected to. Fife (2004: 155) writes that pilgrimage is "fundamentally concerned with the transformative encounter between Self and Other and its potential for spiritual growth and renewal". Although features of the trail like the trees, waterfalls, lakes, mountaintops, and spectacular views all comprised part of this 'Other', it was the animals on the trail who most actively embodied the alterity and opportunity for connection that hiker-pilgrims sought.

In *Biophilia* (1984: 139) Wilson views nonhuman animals as necessary to the lives of humans, describing other organisms as "the matrix in which the human mind originated and is permanently rooted", as offering the "challenge and freedom innately sought", and therefore as a vital part of our lives, which would be incomplete without them. Wilson, like many others, would probably view a pilgrimage centred around encounters with wild animals as the natural reaction to a life deprived of the presence of other autonomous beings. Yet McCabe, writing not specifically about pilgrimage, but about travellers and tourists, describes each type of visitor as looking for essentially the same thing, "a search for the self in the reflection of the other" (2005: 95), taking the focus off of the animal other and placing it firmly back on the self. Many depictions of this search are less than flattering about the traveller, ostensibly because the tourist/traveller is viewed as seeing the animal as a signifier of whatever they are looking for, rather than seeing the animal as themselves (see Bulbeck, 2005; Curtin, 2005; Ingold, 1994; Jarvis, 2000; MacCannell, 1976; Urry, 1990). Curtin cites Kripendorf's use of the term "emotional recreation" to describe journeys like this (2005: 3). For Berger, "the life of a wild animal becomes an ideal...the starting-point of a daydream: a point from which the day-dreamer departs with his back turned" (1980: 17). There is a sense in all of these interpretations that people *use* the wild animal for something that they need, be it to feel connected to the rest of nature, or a desire to "find themselves" out in nature. The implication behind using an animal this way is that we would only be able to see the thing that we need, and not to see the animal as he truly is, to perceive him "directly", as Ingold puts it (1994: 12).

For a few hiker-bloggers, their relationship to the surroundings and lifeforms on the AT was described in terms of a very specific religious connection, with mentions of God, Christianity (nobody explicitly named a religion other than Christianity) and the wilderness as a 'cathedral'. Other people's experience of nature was described in apparently wholly secular terms, as an enjoyment of fresh air, greenery, encounters with interesting animals, and so forth. Yet for many of the people who approached their hike as pilgrimage-like, there was a clear sense that they viewed their wild surroundings as being sacred in some way. Goodnow and Bloom (2017: 11) argue that it is the intentionality with which pilgrims seek out "the sacred" that sets them apart from tourists. With this in mind, and inspired by Durkheim and Eliade, they set out 12 properties of the sacred in order to understand under what conditions travel becomes sacred (and thus a pilgrimage). These 12 properties are: hierophany, kratophany, sacrifice, ritual, opposition to the profane, contamination, commitment, communitas, objectification, ecstasy and flow, myth, and mystery. It is useful to consider these properties when thinking about hiker relationships to the trail and its nonhuman inhabitants.

Hierophany is described as the manifestation of the sacred, or as the sacred showing itself to us. Hikers who described not just delight, but a sense of awe and wonder at seeing a bull moose flinging lake water with his antlers at sunset might be said to be experiencing hierophany. The second property, according to Goodnow and Bloom, is *kratophany*, which describes a dual effect of the sacred, in both drawing individuals in fascination, and simultaneously repelling them in fear, generating an overall sense of power. Certain species, such as black bears, moose, wild boar, snakes, and others were certainly described in terms not just of their individual power to attract and repel at the same time, but also in terms of how their presence made people feel about the trail itself; both deeply at home in the surroundings and intimidated by them. Many of the *sacrifices* made on the trail are carried out in relation to other animals on it; for example, hiker food can be taken away by bears, shelter mice, and other scavengers, and hiker blood can be sucked by ticks and leeches. Similar to sacrifice, *ritual* "establishes rules of conduct regarding how one relates to the sacred" (2017: 11). The Leave No Trace rules advocated by the organisations responsible for the maintenance of the trail, as well being blogged about by many of the hikers on the trail, establish a set of rules that are followed partly in order to ensure the continuing good health of the nonhuman animals on the trail (as well as the human ones). Goodnow and Bloom describe the sacred as being defined partly by *opposition to the profane*, and vitally, by losing its sacred status if "impinged upon by the profane" (2017: 11). There are many examples of how this can happen on the AT, but perhaps one of the most pertinent, which will be explored in the following chapter, is the effect that access to hiker junk food can have on black bears, where hiker food (brought onto the trail from somewhere else) can be viewed as 'profane', bears as 'sacred' (because they belong to the wilderness), and the effect of the food on bears as rendering them 'less sacred'.

The next property of sacredness, *contamination*, represents "the ability of sacredness to be spread via contact" (2017: 11). This is an elemental property of pilgrimage itself, during which the pilgrim hopes for contact with the sacred to have a transformative effect on themselves. Interactions with autonomous animals on the trail formed part of this hoped-for 'contamination' for some hikers. *Commitment* simply describes the state of resolute focus and connection to the perceived sacred. *Communitas* refers to the feeling of camaraderie with fellow participants in the sacred, in this instance, fellow hikers as well as non-human travellers on the trail. Further, the intangible sacred can be *objectified* in a particular object, which then comes to represent the sacred. This object can be a feather, a photograph, or even a trail journal published online.

Ecstasy and flow, viewed together by Goodnow and Bloom as one property of the sacred, are described as producing experiences in which an individual feels joyful and centred, as well as a "loss of the sense of self,", in a description very close to psychologist Csikszentmihalyi's description of flow as a state of "optimal experience" (1990: 3). The final properties of the sacred, *myth* and *mystery*, operate in conjunction with each other in the identification and perpetuation of sacred status. Myth is described as preserving the mystery of the sacred, which is seen as essential for its survival. There is a lot of mythology around the AT, and the lifeforms who dwell there, mythology that is bolstered in part by AT blogger online accounts of their encounters with animals on the trail.

By considering Goodnow and Bloom's proposed properties of the sacred it is possible to conclude that the nonhuman animals on the AT were sometimes viewed, if not as sacred themselves, then certainly as belonging to the sacredness of the wilderness that they inhabited.

Witnessing

One of the recurring motifs in hiker narratives about animals on the trail was the idea of being a 'witness' to their presence. The concept of witnessing can play a significant role in pilgrimage, from witnessing a miracle (see Digance, 2003) or a vision of the Virgin Mary (see Gesler, 1996) to witnessing the field where a battle took place (see West, 2008) or another location where mass killing happened (see Knudsen, 2011, on dark tourism). Trek bloggers, many there to witness 'nature' talked about stopping to observe a mother bear and her cubs climbing a tree, noticing a rabbit hiding in long grass, seeing a doe licking her fawn clean, or a mother bird feeding a baby bird, watching a beaver bobbing in a lake, and a mother moose and her baby swimming across it. One talked about noticing the passing of time on the trail in the birds flying south overhead and the deer growing their winter coats. Another wrote:

> The sun was setting when we arrived as a couple of hikers were gathered at the water's edge, pointing excitedly. We squinted against the sun's glare

and there, silhouetted perfectly, was a massive bull moose, flinging the water with his antlers.

In a blog post addressed to the state of Maine, a hiker wrote "You showed me wildlife: squirrels, snakes, snowshoe hares and more". Others emphasised the movement of animals on and around the trail.

> As we made our way back to the trail, we stood by the side of the road and watched a herd of deer run across into the huge open expanse of Big Meadows. They were jumping higher than I've ever seen, cutting through the fields with ease. It was like watching dolphins seamlessly pushing through an ocean current.
>
> We had a falcon fly right in front of us! It was soaring and tucked its wings to dive as it passed. Amazing! One night while cooking dinner, a mole came out of a hole, looped around us, and entered a different hole. To our surprise, it came out of the first hole again, and again! Or maybe it was a mole migration!

In these types of narrative, the animal, if aware of the hiker's presence at all, did not mind the hiker stopping to observe their activities. One person said, "I stopped to gaze at a black snake...it apparently paid no mind as I got within half a foot of it to simply sit and watch it taste the air, looking for prey". Thus, the hiker could portray himself as fitting in with the animals of the trail, who knew that he was not a threat to them, and therefore allowed him to bear witness as they went about their lives. Barron (1875: 324) wrote that "there are times when it is good for a man to walk alone; nature has her privacies, and won't reveal them to you nor me when others are listening".

Hikers could further portray themselves as more sensitive to the other creatures of the trail than a 'tourist' might be, as unlike in most ecotourism narratives, encounters with small, even tiny, animals were depicted as equally important to the observer as encounters with large animals. Bloggers talked about stopping to notice tadpoles swimming in a puddle, coming across a "curious field mouse", watching as a spider "literally came out of a water-spout" or weaving a web, a single butterfly in a meadow, noticing the "beautiful colors" of salamanders, and admiring the delicateness of moths.

Bulbeck describes her trip to Antarctica, saying, "There were so few of us and so many of them that humans felt peripheral" (2005: 151).

> The length of time that separated us from 'civilization' and our puny numbers in the landscape meant that we saw ourselves briefly as the environment saw us. In a sense, the environment did not see us. The animals paid us no mind and we humans had been "rendered absolutely marginal", to use Berger's (1980, p. 22) phrase.
>
> (Bulbeck, 2005: 151)

Far from being discomfiting, this feeling of marginality seems to have been relished by many of the people who blogged about hiking the trail, in line with the idea of wanting to feel the presence of something 'bigger' than themselves (see Bratton, 2012: 21).

Walking in a woodland wonderland

In narratives centred around witnessing animal life on the trail, the animal was depicted as carrying on their life-ways unperturbed by the presence of the hiker. Other blog posts described animals unworried by the presence of the human, but noticeably aware of them; these posts tended to describe the hiker and animal as harmoniously sharing the space together. Often this included the implication that this harmonious meeting would not happen outside of the trail. One hiker wrote that, "I sat eating lunch trailside and an eight-point buck in velvet walked up. We ate lunch beside each other, glancing up once in a while". His use of hunting terminology is juxtaposed with the image of him and the buck peacefully eating together, reinforcing the notion of the trail as sacred space. Another blogger described feeling very depressed until a bird landed near to him and "sang to me for several minutes", after which all of his worries dissipated. Many of the narratives about harmonious encounters featured the birds of the trail.

> A hawk followed me down the trail for a few minutes, making short flights from tree to tree…I stopped at a rocky outcropping to watch some turkey vultures flying around; they were gliding past me on updrafts, so close I felt like I could reach out and brush my fingers against their wing tips. I felt like I could have joined them.

Like Bulbeck describing her trip to Antarctica, these blogs described being in "communion" with the other animals on the trail, rather than in "communication" (2005: 149) with them, or as Abram writes about being a human in the spaces of non-humans, "there is an intimacy here that includes you" (2011: 20).

This feeling of intimacy often led to descriptions of encounters as 'magical', or even of the trail as a kind of wonderland where people were woken by birds chirping in the treetops above, and went to bed listening to owls hooting and the "gentle quacking of ducks sleepily going to bed". One hiker wrote that, "last night I walked into a wood line of fireflies. For a few moments I had my own little wonderland". Another described her hiking partner's encounter with a chipmunk.

> He had picked up a pine branch to use as a walking stick for his knee that morning, and joked that it did not, contrary to all expectations, make him feel more like a wizard. He promptly left it behind at the shelter, and had

to run back to get it. When he walked back towards me, he flourished it like a staff, and a chipmunk burst from the ground at his feet like he had shot it out of his walking stick. And that's how he was named Radagast, after the wizard in Tolkien's works who has such an affinity with wild animals.

One blogger wrote about fawns coming within a few feet of him, bears wandering calmly through campgrounds and birds that sat and watched him walk by, concluding that the animals of the trail were putting on a "quasi Snow White show". Later, the same blogger described his experience of coming across ponies.

> My first view of the ponies was when I was coming down from Thomas Knobb. I crossed over some rock formations and began to feel like I was walking through some kind of faerie wonderland. Open grass fields spotted with fantastic boulders, rhododendrons in bloom, petals falling to the ground carpeting the trail like some kind of romantic welcome. Then I turned a corner and spotted the first herd.
> They were running and playing, manes flowing in the wind.

Interactions with animals were also frequently described as magical. People described feeding ducks by hand and being licked by ponies. A mouse tickled one hiker's beard with his whiskers, and another hiker was woken by a pony wiggling her nose into his tent and nuzzling him. Someone else stopped for a night at a hostel, where the manager had chipmunks running around her feet, and eating almonds out of her hands. Disneyesque experiences were written about with pleasure.

> In a somewhat nondescript portion of the trail, Magnus noticed a bunch of Gray Jays kind of following us. Having heard of their familiarity with hikers, he held out his hand. A bird landed on it!!! It was like living in a Disney movie.

Another blogger wrote, tongue in cheek, about how thru-hikers eventually learned to communicate with the animals on the trail, saying, "as many people know, once you pass Fontana Dam, your powers allowing you to talk to animals kick in". He listed the advantages to this, which included being able to talk bears into stealing other people's picnics for you, and getting sparrows to do your laundry for you, like "Snow freaking White".

Alexander Wilson (1992: 128) writes that "the 'lifeforce' in the core of the apple that Disney tried so hard to film must finally be the life in our own human bodies, which are inextricably connected to the rest of the biophysical world". Returning to the idea of pilgrimage as a quest for connection, or as Wilson writes, "a deep desire simply to be in the world" (1992: 128), hiker narratives describing encounters with animals on the trail as magical or Disney-like centred around the feeling of being accepted by the animal community living there, or even feeling *welcomed* by them. As Candea and da Col write, "hospitality is magic" (2012: s3). They describe hospitality as a fascinating concept due to the "dangers embedded in any encounter" (2012: s3), contrasted with "the necessary ethical requirement of absolute openness to the Other" (2012: s4). Citing the ideas of Pitt-Rivers (1968) and Derrida, (2000) they continue: "the stranger is the absolute unknown, whose radical alterity echoes the numinous presence of the divine itself" (2012: s6). Thus, the hiker's desire for feelings of 'connection' was requited by encounters with animals on the trail that made them feel welcomed, and the idea of Nature as the source of sacredness and connection was reinforced.

Conclusion

Undoubtedly, a hiker's sense of why she is on the trail and what she hopes to achieve there meaningfully influences her perception and experience of the animals on the trail. One of the ways that a thru-hike of the AT can be approached is as a pilgrimage, and encounters with animals on the trail are often interpreted by hikers through this notion of being on a pilgrimage out in the wilderness. Of course, the extent to which these ideas will apply to each hiker will depend upon the individual, and can vary throughout the five to seven months that they are hiking (see Cohen, 1979; McCabe, 2005). Ingold questions whether animals exist for us as meaningful entities only through the symbolic values that they hold for us, or whether we are able to experience them "directly" through our common immersion in a shared environment (1994: 12). Carman (2014: xiv) writes that "perception is always both passive and active, situational and practical, conditioned and free". Hiker experiences of animals as symbols of their pilgrimage were situational and conditioned, while they were also able to engage with those animals in ways both practical and free. Often it was clear that encounters were interpreted as encouraging signs that what was being sought – freedom, authenticity, connection – was being achieved. Sometimes hikers indicated that it was the perceived sacred space of the trail itself that made this possible –

> I've seen thousands of deer in my lifetime. They're all over…where I grew up. It's totally different when I'm out on the trail. I look at them as majestic creatures instead of the pests I was taught they were growing up

Yet hikers witnessed characteristics and behaviours in other animals that they might never have known about, and had contact – even communication – with animals, that they might never have thought possible. They were often delighted to report the activities of a moose, a frog, a raccoon or even a skunk, that they had engaged with. As with the narrator who described his different attitude to deer on the trail, for many of these hikers it may have been the first time that they had stopped to pay attention to a frog or a bird, given that as a pilgrim it was their 'job' to pay attention, in order to achieve the 'connection' explicitly sought. Thus, the pilgrimage perspective of an AT hike has the potential to draw the pilgrim-hiker's attention to the personhood of the other animals around them. Rather than seeing animals as "signifiers" of a certain "folk taxonomy" (Ingold, 1994: 12), the focus on the animal that is encouraged by the notion of being out on the trail to find radical otherness actually encouraged the hiker to experience the animal in a more intimate and personal way, even as the hiker simultaneously experienced encounters with animals as 'evidence' of achieving their pilgrimage goals. In *Being Alive* (2011) Ingold cites Gibson's insistence (1979) that perception is the achievement of an organism as it moves about in its environment, and that "what it perceives are not things as such but what they afford for the pursuance of its current activity" (2011: 11), yet that "it is in the very process of these 'affordances'...in the course of their engagements with them, that skilled practitioners – human or non-human – get to know them" (2011: 11).

References

Abram, D. 2011. *Becoming Animal*. New York: Random House.

ALDHA Facebook Group. 2015. Available at: www.facebook.com/groups/aldha/

Badone, E. 2004. Crossing Boundaries: Exploring the Borderlands of Ethnography, Tourism and Pilgrimage. In: E. Badone and S. R. Roseman (eds.) *Intersecting Journeys: The Anthropology of Pilgrimage and Tourism*, pp. 180–190. Chicago: University of Illinois Press.

Badone, E. and Roseman, S. R. 2004. Approaches to the Anthropology of Pilgrimage and Tourism. In: E. Badone and S. R. Roseman (eds.) *Intersecting Journeys: The Anthropology of Pilgrimage and Tourism*, pp. 1–23. Chicago: University of Illinois Press.

Baron, D. 2004. *The Beast in the Garden*. New York: W. W. Norton.

Barron, A. 1875. *Footnotes, or Walking as a Fine Art*. Connecticut: Wallingford Printing Company.

Berger, J. 1980. *About Looking*. London: Bloomsbury.

Bratton, S. P. 2012. *The Spirit of the Appalachian Trail: Community, Environment, and Belief on a long-distance hiking path*. Knoxville: The University of Tennessee Press.

Bulbeck, C. 2005. *Facing the Wild*. London: Earthscan.

Candea, M. and da Col, G. 2012. Introduction: The Return to Hospitality. *Journal of the Royal Anthropological Institute* 18(S1): S1–S19.

Carman, T. 2014. Foreword. In: M. Merleau-Ponty *Phenomenology of Perception*, pp. vii–xvi. New York: Routledge.

Cohen, E. 1979. A Phenomenology of Tourist Experience. *Sociology* 2(13): 179–201.

Coleman, S. and Elsner, J. 1995. *Pilgrimage: Past and Present in the World Religions*. Cambridge: Harvard University Press.

Collins-Kreiner, N. 2016. The Lifecycle of Concepts: The Case of 'Pilgrimage Tourism'. *Tourism Geographies* 18(3): 322–334.

Collins-Kreiner, N. and Kliot, N. 2000. Pilgrimage in the Holy Land: The Behavioural Characteristics of Christian Pilgrims. *GeoJournal* 501: 55–67.

Cronon, W. 1996. The Trouble with Wilderness; or, Getting Back to the Wrong Nature. In: W. Cronon (ed.) *Uncommon Ground: Rethinking the Human Place in Nature*, pp. 69–90. New York: W. W. Norton.

Csikszentmihalyi, M. 1990. *Flow: The Psychology of Optimal Experience*. New York: Harper Collins.

Curtin, S. 2005. Nature, Wild Animals and Tourism: An Experiential View. *Journal of Ecotourism* 4(1): 1–15.

Darby, W. J. 2000. *Landscape and Identity*. New York: Berg.

Derrida, J. 2000. *Of Hospitality*, trans. R. Bowlby. Stanford: Stanford University Press.

Digance, J. 2003. Pilgrimage at Contested Sites. *Annals of Tourism Research* 30(1): 143–159.

Edensor, T. 2000. Walking in the British Countryside: Reflexivity, Embodied Practices and Ways to Escape. *Body & Society* 6(3–4): 81–106.

Evernden, N. 1992. *The Social Creation of Nature*. London: The Johns Hopkins University Press.

Fife, W. 2004. Extending the Metaphor: British Missionaries as Pilgrims in New Guinea. In E. Badone and S. R. Roseman (eds.) *Intersecting Journeys: The Anthropology of Pilgrimage and Tourism*, pp. 140–159. Chicago: University of Illinois.

Fondren, K. M. 2016. *Walking on the Wild Side*. New Jersey: Rutgers University Press.

Frey, N. L. 2004. Stories of the Return: Pilgrimage and Its Aftermaths. In: E. Badone and S. R. Roseman (eds.) *Intersecting Journeys: The Anthropology of Pilgrimage and Tourism*, pp. 89–109. Chicago: University of Illinois.

Fullagar, S. 2000. Desiring Nature: Identity and Becoming in Narratives of Travel. *Cultural Values* 4(1): 58–76.

Gesler, W. 1996. Lourdes: Healing in a Place of Pilgrimage. *Health & Place* 2(2): 95–105.

Gibson, J. J. 1979. *The Ecological Approach to Visual Perception*. Boston: Houghton Mifflin.

Goodnow, J. and Bloom, K. S. 2017. When Is a Journey Sacred? Exploring Twelve Properties of the Sacred. *International Journal of Religious Tourism and Pilgrimage* 5(4): 10–16.

Holmes-Rodman, P. E. 2004. "They Told What Happened on the Road": Narrative and the Construction of Experiential Knowledge on the Pilgrimage to Chimayo, New Mexico. In: E. Badone and S. R. Roseman (eds.) *Intersecting Journeys: The Anthropology of Pilgrimage and Tourism*, pp. 24–51. Chicago: University of Illinois.

Ingold, T. 1994. Preface to the Paperback Edition. In: T. Ingold (ed.) *What Is an Animal?*, pp. xix–xxiv. London: Routledge.

Ingold, T. 2011. *Being Alive*. London: Routledge.

Isenberg, A. C. 2002. The Moral Ecology of Wildlife. In: N. Rothfels (ed.) *Representing Animals*, pp. 48–64. Bloomington: Indiana University Press.

Jarvis, C. H. 2000. If Descartes swam with dolphins: the framing and consumption of marine animals in contemporary Australian tourism. Unpublished PhD thesis, Department of Geography and Environmental Studies, University of Melbourne, Melbourne.

Knox, D. and Hannam, K. 2015. The Secular Pilgrim: Are We Flogging a Dead Metaphor? In: T. V. Singh (ed.) *Challenges in Tourism Research*, pp. 46–52. Bristol: Channel View Publications.

Knudsen, B. T. 2011. Thanatourism: Witnessing Difficult Pasts. *Tourist Studies* 11(1): 55–72.

Lois Gonzalez, R. C. 2013. The Camino de Santiago and Its Contemporary Renewal: Pilgrims, Tourists and Territorial Identities. *Culture and Religion* 14(1): 8–22.

MacCannell, D. 1973. Staged Authenticity: Arrangements of Social Space in Tourist Settings. *American Journal of Sociology* 793: 589–603.

MacCannell, D. 1976. *The Tourist: A New Theory of the Leisure Class.* New York: Schocken Books.

McCabe, S. 2005. "Who Is a Tourist?" A Critical Review. *Tourist Studies* 5(1): 85–106.

McKibben, B. 2003. *The End of Nature.* London: Bloomsbury.

Merleau-Ponty, M. 2014 [1945]. *Phenomenology of Perception.* Oxon: Routledge.

Morinis, E. A. 1992. Introduction: The Territory of the Anthropology of Pilgrimage. In: E. A. Morinis (ed.) *Sacred Journeys: The Anthropology of Pilgrimage*, pp. 1–27. Westport: Greenwood.

Nash, R. 1982. *Wilderness and The American Mind.* New Haven: Yale University Press.

Nolan, M. L. and Nolan, S. 1992. *Christian Pilgrimage in Modern Western Europe.* Chapel Hill: University of North Carolina Press.

Oelschlaeger, M. 1991. *The Idea of Wilderness.* New Haven: Yale University Press.

Olwig, K. 1993. Sexual Cosmology: Nation and Landscape at the Conceptual Interstices of Nature and Culture; or What Does Landscape Really Mean? In: B. Bender (ed.) *Landscape: Politics and Perspectives*, pp. 307–343. Oxford: Berg.

Pitt-Rivers, J. A. 1968. The Stranger, the Guest and the Hostile Host: Introduction to the Study of the Laws of Hospitality. In: J. G. Peristiany and J. Pitt-Rivers (eds.) *Contributions to Mediterranean Sociology: Mediterranean Rural Communities and Social Change*, pp. 13–30. Paris: Mouton.

Porter, J. E. 2004. Pilgrimage and the IDIC Ethic: Exploring *Star Trek* Convention Attendance as Pilgrimage. In: E. Badone and S. R. Roseman (eds.) *Intersecting Journeys: The Anthropology of Pilgrimage and Tourism*, pp. 160–179. Chicago: University of Illinois Press.

Price, J. 1999. *Flight Maps.* New York: Basic Books.

Raglon, R. 2009. The Post Natural Wilderness and Its Writers. *Journal of Eco-Criticism* 1(1): 60–66.

Rinschede, G. 1992. Forms of Religious Tourism. *Annals of Tourism Research* 19: 51–67.

Roseman, S. R. 2004. Santiago de Compostela in the Year 2000: From Religious Center to European City of Culture. In: E. Badone and S. R. Roseman (eds.) *Intersecting Journeys: The Anthropology of Pilgrimage and Tourism*, pp. 68–88. Chicago: University of Illinois Press.

Salleh, A. 1996. The Politics of Wilderness: Aborigines and Eco-Activists. *Arena Magazine* 23: 26–30.

Sallnow, M. J. and Eade, J. (eds.) 1991. *Contesting the Sacred: The Anthropology of Christian Pilgrimage.* London: Routledge.

Schänzel, H. A. and McIntosh, A. J. 2000. An Insight into the Personal and Emotive Context of Wildlife Viewing at the Penguin Place, Otago Peninsula, New Zealand. *Journal of Sustainable Tourism* 8(1): 36–52.

Scriven, R. 2014. Geographies of Pilgrimage: Meaningful Movements and Embodied Mobilities. *Geography Compass* 8(4): 249–261.

Terry, D. P. and Vartabedian, S. 2013. Alone but Together: Eminent Performance on the Appalachian Trail. *Text and Performance Quarterly* 33(4): 344–360.

Thoreau, H. D. 2007 [1861]. *Walking.* Rockville: Arc Manor.

Timothy, D. J. and Olsen, D. H. (eds.) 2006. *Tourism, Religion and Spiritual Journeys.* New York: Routledge.

Tuan, Y. 1998. *Escapism.* London: The Johns Hopkins University Press.

Turner, V. and Turner, E. 1978. *Image and Pilgrimage in Christian Culture.* New York: Colombia University Press.

Urry, J. 1990. *The Tourist Gaze.* London: SAGE.

Vukonic, B. 1996. *Tourism and Religion.* Oxford: Elsevier.

Washington, H. G. 2007. The "Wilderness Knot". *USDA Forest Service Proceedings* RMRS-P-49: 441–446.

West, B. 2008. Enchanting Pasts: The Role of International Civil Religious Pilgrimage in Reimaging National Collective Memory. *Sociological Theory* 26(3): 258–270.

West, P. and Carrier, J. G. 2004. Ecotourism and Authenticity: Getting Away from It All? *Current Anthropology* 45(4): 483–498.

Whatmore, S. and Thorne, L. 1998. Wild(er)ness: Reconfiguring the Geographies of Wildlife. *Transactions of the Institute of British Geographers* New Series 23(4): 435–454.

Wilson, A. 1992. *The Culture of Nature.* Oxford: Blackwell Publishers.

Wilson, E. O. 1984. *Biophilia.* Harvard University Press.

2 'Real' wilderness needs a 'real' predator
Hikers and black bears

The black bear is the only member of the family *Ursidae* to inhabit the Eastern United States (Adkins, 2000). They occupy every state along the AT, but are perhaps most abundant in certain areas such as the Great Smoky Mountains, Shenandoah National Park, the Pennsylvanian state forests, and the Maine woods (Adkins, 2000). On some sections of the AT bears are protected, and on other sections they are hunted.

Black bears have typically been characterised as powerful and dangerous animals (see Pelton et al., 1976); as a species they are one of the quintessential charismatic megafauna. They are highly intelligent and extremely adaptable, have been acknowledged as having great plasticity in their relationships with conspecifics, and also possess a high level of adaptability in their contact with humans (Tate and Pelton, 1983). Their adaptability, as well as their natural curiosity, can be seen in their opportunistic food-obtaining behaviours around people, which often lead to their becoming habituated to being around human beings in order to gain access to their food (Clark et al., 2002), in turn leading to 'nuisance' behaviours. However, black bears are naturally shy animals and have commonly tended to avoid human beings and developed areas, and it has been argued that foregoing these natural tendencies in order to approach people produces stress (see Clark et al., 2002; Tate and Pelton, 1983). This may be why they tend to enter developed areas (for example a camp site) during the night, when there is less human activity evident.

There have been numerous studies conducted on black bear behaviour in the Appalachian region (see Carney and Vaughan, 1987; Clark et al., 2003, 2005; Eagle and Pelton, 1983; Garshelis and Pelton, 1980, 1981; Mitchell and Powell, 2003; Pelton, 1972; Powell and Seaman, 1990; Reynolds-Hogland and Mitchell, 2007; Reynolds-Hogland et al., 2007; Williamson and Whelan, 1983; Yarkovich et al., 2011), particularly with regard to roaming and feeding behaviours, and habitat selection. Studies have also looked specifically at interactions between black bears and humans (Singer and Bratton, 1980; Tate and Pelton, 1983) and at human attitudes towards black bears (Baptise et al., 1979; Burghardt et al., 1972; Gore et al., 2006; Pelton et al., 1976).

Much of this research commenced in the 1970s and 1980s. In 1972 Burghardt et al. looked at knowledge and attitudes concerning black bears by visitors to the Great Smoky Mountains National Park. They found that the close proximity of bears and people in the national park was affecting the habits of many, if not most, of the bears dwelling there, and many had become habituated to humans, leading to them being labelled as 'nuisance bears'; behaviour which frequently originated with humans actively feeding the bears and encouraging them to come closer. At the time, the park had a policy of trying to discourage nuisance bears (bears who approached people for food) from 'reoffending' by tapping them on the nose with a baseball bat (1972: 259). They tried relocating bears to other areas of the park, but this strategy failed, largely due to the bears' rehoming abilities (Burghardt et al., 1972: 269).

In 1976 Pelton et al. studied the attitudes of people who had experienced property damage or injury by a black bear in the Smokies. They too found that the urge to feed bears was very strong in people, attributing it to the bears' "anthropomorphic appearance" (Pelton et al., 1976: 163). Most 'victims' were found not to have followed advice regarding avoiding conflict with a bear. Many of these people were quick to want the bear involved to be punished in some way, but these attitudes seemed to change over time, and many eventually conceded that the fault had been their own (Pelton et al., 1976: 164–165). Similarly, Baptiste et al.'s 1979 study was conducted to try to determine the public acceptance level for noninjurious human–black bear contact at Shenandoah National Park. They found that most people were willing to acknowledge that visitor carelessness was causing the majority of bear "problems" (1979: 25).

Singer and Bratton (1980) found that by the late 1970s many human–black bear conflicts were already occurring at campsites and shelters along and around the AT in the Smoky Mountains. Their paper concluded that "a challenge to park management will be to redistribute or even minimise the bear/human conflicts resulting from ever-expanding visitor use along this trail" (1980: 139). In 1983 Tate and Pelton found that people were still feeding bears, treating them like "household pets" (1983: 319). They warned that, "while bears may be exploiting a new food resource; they are not seeking association with humans; they are wild animals" (1983: 317). Accordingly, agonistic behaviour in bears was usually triggered by the actions of people, the most common precipitator of aggression being crowding by visitors (Tate and Pelton, 1983: 314).

More than 20 years later, Gore et al. (2006) conducted a study into visitor perceptions of risk associated with human–black bear conflicts in New York's Adirondack Park, which the AT runs through. Reports of non-lethal encounters were frequent, and included minor scratches, lost backpacks and food, ripped tents, and broken car windows (2006: 36). Despite incidences of human–bear conflict being relatively high in campgrounds in the park, overall perceived risk was low. It was found that public perception of bears was influenced by bears' phylogenetic similarity to people, their high intelligence, their aesthetic appeal, their relatively large size and capacity to stand erect, their omnivorous diet, and their historical and cultural relationship with people (Gore et al., 2006: 37). As with the previous studies mentioned, the success of people and black bears coexisting was acknowledged as dependent as much on human behaviour as on bear 'management'.

Yet close proximity to humans has been shown to be disruptive to bears in multiple ways, producing stress (Tate and Pelton, 1983) and changing their natural habits (Burghardt et al., 1972). Reynolds-Hogland and Mitchell (2007) found that bears would avoid areas near to gravel roads, and Reynolds-Hogland et al. (2007) cited several studies showing that when den sites are in close proximity to humans there is an increase in overwinter weight loss and abandonment of cubs. Rogers (1987) found that human disturbance may be one of the most important influences on where black bears select their den sites. When bears become habituated to humans and seek out their food there is the strong possibility of conflict, and to the bear being labelled as a 'nuisance' bear, and either being captured and moved from his home-site, being captured and released on-site (the idea being that the bear is traumatised from the experience, leading him to avoid further contact with humans, see Clark et al., 2002), or captured and killed. If the bear is not punished for seeking out human food, there is the possibility of other side effects of subsisting on food produced for humans, such as the suggestion that the bear's reproductive capacity can be detrimentally affected by this diet (see Cole, 1974; Stokes, 1970). Even when a bear is not necessarily a 'nuisance' bear his presence in the vicinity of a nuisance bear can lead to him being mistaken for that bear and killed, as in a 2016 case at Spence Field (WNCN, 2016).

On 10 May 2016, AT hiker Bradley Veeder pitched his tent near to the Spence Field Shelter in the Great Smoky Mountains National Park (Veeder, 2016). A couple of hours after falling asleep he was woken by a bear biting into his right leg. The bear had ripped a large hole in Veeder's tent, and over the next few minutes attacked and retreated several times over, as Veeder fought the bear off by punching and shouting at him. During a ten-minute interval Veeder took a chance and limped to the nearby shelter (shelters on the AT tend to be basic structures formed by three walls and a roof, with the final side open to the elements), where another hiker applied first aid to the puncture wounds in his leg. In the morning it was found that the bear had "chewed everything that I left behind (tent, tent poles, backpack, water filter, water bottles, phone, book, etc). With extremely lucky timing, I had made it to the shelter after the initial attack and before the bear returned" (Veeder, 2016). A few days later National Park Service (NPS) wildlife staff killed an adult male bear who was spotted near to the Spence Field shelter, and was initially believed to be the bear who had attacked Bradley Veeder. After performing DNA analysis on the euthanised bear it was found that the bear's DNA was not a match to the DNA of Veeder's attacker (WNCN, 2016).

Approximately a fifth of the hikers reviewed reported having had at least one interaction with a black bear. If we take the Trek bloggers as a roughly representative sample of all long-distance hikers on the AT during 2015 and 2016, bearing in mind that the possibility that not everyone who had an inter-action wrote about it, it is possible to surmise that around 20 percent of hikers had encounters with bears. Owing to the very high volume of hikers on the AT during the spring to autumn months each year, this amounts to poten-tially thousands of interactions between humans and bears, most of which may have ended swiftly (the bear runs away or the human retreats), but many of which will have involved some level of conflict, including stand-offs, camp invasions, and bluff charges. With more and more people undertaking hikes on the AT every year, the potential for conflict rises. Hiker-bloggers' accounts of dwelling among and sometimes interacting with bears on and around the AT can contribute to an understanding of how humans and bears perceive each other and find ways of co-existing together, and what happens when they get 'too close for comfort'.

Of 1,691 blog posts reviewed, 90 posts were found to feature stories about, or making significant mention of, a bear or bears. All blogs containing references to bears were coded according to type. In the first type, hiker-bloggers talked about bears in general, rather than a specific incident. Posts covered topics such as how a novice hiker was preparing for a possible encounter with a bear, or advice from a veteran long-distance hiker on how to behave during a black bear interaction in order to minimise the risk of getting hurt. Most of this type of post described what people expected about bears. The second type of post described the co-existence of hikers and bears on the trail. This covered a hiker-blogger's awareness of the presence of bears whilst on the trail, and often made mention of incidents such as a bear sighting or having heard bear-like sounds in the night. In the third type of post, bloggers described specific encounters or interactions that they had with a bear or bears whilst on or near to the AT. Occasionally a blog would be coded with more than one type, if for example it described two separate events, one an incident alerting the hiker to bear presence in the area and the other a direct encounter with the bear.

Analysis of blog posts focused on hiker perceptions of bears, how that affected how they behaved around and towards bears, how bears behaved around and towards hikers, and how their behaviour was interpreted by hikers. It is also interesting to consider how hiker-bloggers chose to construct their narratives about bears, including not only how they portrayed bears, but how they portrayed themselves interacting with bears. Stories are, among other things, ways of self-making, and bears were often used as a tool for the narrator to say something about themselves.

Anticipating bears!

Bloggers who had not yet started their hike frequently blogged about what they expected from the long-distance hike experience. The bloggers antici-pating starting their hike usually included bears in narratives about what they were most scared of, as in a post entitled "Fear is the mind-killer", in which bears are at the top of the blogger's list of fears.

> So in the dark, picturing myself in a tent and one of these giant toothsome beasts rummaging around outside for a crumb of the candy bar that's probably still stuck to my tee-shirt, makes the blood run a bit cold. I know about the precautions and safeguards, and I know that I'll take them seriously, at least at first. What happens after one of those long, exhausting days that I keep reading about and I forget I jammed some cookies in my backpack before I put it in my tent?

Around a quarter of blog posts anticipating bears characterised the possibility as threatening, whereas another quarter of posts about the potential danger posed by an interaction with a bear, written both by novice AT hikers and experienced ones, emphasised having respect for bears as opposed to being afraid of them. A few writers quoted friends or family members who had asked them if they weren't afraid of being attacked by a bear while out on the trail, responding that their attitude would be one of respect, not of fear ("bears…are large powerful creatures and should be treated with the utmost respect"). A common trend throughout hiker blogs about bears (whether it was anticipating them, dwelling among them or interacting directly with them) was that hikers and aspiring hikers were keen to demonstrate that they had a knowledge and understanding of black bears. Bloggers advised other hikers not to "sneak up" on bears, to make their presence known "with a pleasant 'Hey Bear'", and to give them "plenty of space, moving off the trail if necessary". One writer, in a post advising hikers to use bear canisters to store their food overnight (so that the bear can't smell the food and be attracted to the camp), pointed out that the canisters were there not just for the safety of hikers, but for the bear's safety as well, saying that "Canisters are just a more effective means of keeping human food away from any wildlife. If animals come in contact with a hiker's food, the 'wild' in 'wilderness' is compromised". Another writer posted that he was respectful of bears, not afraid of them, and therefore would not be taking any bear spray with him on his hike. The safety of the bears themselves was mentioned again in an extensive post by a veteran thru-hiker, who mentioned the many bears who had been "needlessly killed" in her hometown following interactions with hikers and locals who did not follow the correct precautions. Her advice included instructions to never approach a bear, not to run from a bear when encountering one, to try to scare the bear away by making loud noises and creating a large physical presence, to back away slowly and give him space should he appear aggressive, never to drop food around a bear, never to climb up a tree to try to escape a bear, and to fight back if attacked (her advice is in line with the advice of the US NPS on behaviour around American Black Bears (NPS, 2016) – there is different advice for humans interacting with Grizzly Bears, who will respond in a different manner). Overall, bloggers appeared knowledgeable, or at least interested, in bear behaviour, with more than half of the people who blogged about bears demonstrating some level of understanding of bear characteristics and/or the correct way to deal with an encounter. Indeed, these bear advice narratives were frequently used by bloggers throughout the Trek blogosphere (including both experienced and novice hikers) as a way of positioning themselves as authority figures on wilderness hiking in general. As in the comment regarding human-habituated bears having their wildness 'compromised', black bears often seemed to be regarded as metonyms for wildness on the AT, and knowing how to deal with a bear meant knowing how to handle yourself in the wild.

As a counter-narrative to the earnest advisory posts, several bloggers used humour in talking about potentially dangerous scenarios that might arise in an encounter with a bear, often parodying official advice with 'advice' of their own.

> Studies show the best way to deal with a bear is to kick them square in the testicles. If that fails, well...there's a reason only 20% of prospective thru-hikers make it to Katahdin. Seriously though, I totally respect bears. They're huge, and according to Winnie the Pooh, they are capable of pulling some Exorcist-level shit when provoked. For that reason, I will definitely not be provoking them.

Hiker-bloggers also used humour when imagining what might happen to them as a result of a bear encounter. One writer quoted her friends as saying "You're going into the bear's house...so basically he's going to eat you", responding that "I seriously better not get attacked by a bear just so I don't have to hear it from my friends". Another pictured himself in a boxing match with a bear.

> If you ever find yourself face-to-face with one of these beautiful creatures, you need to alert the bear that you are not to be fucked with. What to do: Assert dominance by squaring up with the bear; the most alpha thing you can do is to show it you're willing to go toe-to-toe in a fist fight. Put up your dukes and start bobbing and weaving towards the animal. If you find that the bear hasn't turned to run by the time you're within striking distance (unlikely), give it a quick one-two jab in the stomach to show it you mean business. Bears are big, but are notorious for being poor boxers, so more often than not they will run when propositioned to a fight.

There was evidence that some, if not most, people had conflicting feelings about a possible encounter with a bear. One blogger mentions the "numerous black bears" that "we both did and did not want to see". Another, about midway through his hike, comments that he has not seen any bears yet, making him either lucky or unlucky, "depending on who you ask". Others were frank about their disappointment at not having seen a bear yet.

> Our chatter and noisy steps were probably deterring any wildlife within seeing distance, but Elm was disappointed. "They say you can't help but see a bear through here and we haven't seen anything!" he complained. "We haven't even seen a deer!" And on the word deer, a doe who must have been bedded down right next to the trail shot up, took one look at us, and ran in the opposite direction. "We haven't even seen a bear!" Elm tried, but without success.

Overall, posts written by hikers anticipating bears varied in tone, from serious and respectful to light-hearted and jocular. Yet in spite of the tone, whether earnest or full of self-deprecating bravado, there is a frisson of excitement that comes across in most of these blogs thinking about bears, despite the fact that only one person openly stated that they were excited about the prospect of seeing one (nobody said that they definitely did not want to see one). It is also worth noting again that for many hikers, encountering wild animals is not the primary motivator for their hike, and for several, it may not be a factor at all. Despite this, there certainly seemed to be an implicit understanding among expectant hikers that hiking the entire AT without at least glimpsing a black bear would render the experience somehow incomplete.

> "Aren't you afraid of bears?" It's an oft asked follow-up question upon learning I'm hiking the Appalachian Trail. And rightfully so.
> Within half an hour of being in New Jersey we saw two bears... Quintessential AT day complete.

Expectations about bears: authentic wildness, authentic wilderness

The Trek blogger narratives built up a picture of an animal that is at once ferocious and funny, cuddly and deadly, but above all, *wild*. When thinking about bears before starting their hikes, many bloggers anticipated that they would at some point encounter a bear face-to-face and all appeared to acknowledge that this would be a dangerous situation. In telling how these interactions might go, they tended towards one of three types of narrative. In the first, bloggers spoke openly of their fear about meeting a so-called 'ferocious predator' in the wild. In the second, bloggers were able to show knowledge and authority regarding bears, when detailing how they would approach a bear interaction with an attitude of respect, and not fear. Narratives that demonstrated their competence and understanding of bear behaviour or characteristics – and by implicit extension, knowledge and competence about living in the wild – could be seen as, amongst other things, either subtle or overt attempts at generating what Bourdieu would term embodied cultural capital (Bourdieu, 1986), capital that can be built through the attainment and display of knowledge within a specific social field. Given the assumption that the audience for a blogger's posts would be other AT hikers or aspiring AT hikers, displays of competence in this setting could understandably be expected to endow cultural capital on the writer, in the form of 'someone who knows how to handle themselves in the wild'.

In the third type of narrative, bloggers created a comical counter-narrative, using humour to describe a fish-out-of-water situation in which either the bear would get the better of them, or they would better the bear. Beeman (1999: 103) writes that the most common basis for humour is surprise, and the most common kind of surprise comes under "the general rubric of 'incongruity'". The author of the blog about frightening a bear off by engaging in a boxing match with him invoked incongruity – even absurdity – to create the humour in his narrative. One possible reading for this type of irreverent narrative is as a reaction against what might be viewed by the blogger as an overabundance of earnest advice on how they should deal with a bear, or the 'correct' way to disengage from a bear encounter, advice which might be viewed by some as condescending and deserving of parody. Nevertheless, even these humorous counter-narratives can be seen to be attempts at generating cultural capital, as one of the common traits that can be seen to recur in many of the people who choose to hike the AT is a desire to escape a society they view as too full of rules and regulations.

What is clear from the narratives anticipating bear encounters is that bears were considered as metonymic for wilderness, and for the AT itself. The presence of bears on and around the trail was considered as vouching for the authenticity of the wilderness experience. In their paper "Towards a Conceptual Framework for Wildlife Tourism" (2001), Reynolds and Braithwaite cite a report prepared for Alberta Tourism, featuring a list of suggested characteristics of animals that would make good subjects for wildlife-based tourism. These animals should be: predictable in activity or location, approachable, readily viewable (open habitats), tolerant of human intrusion, possess elements of rarity or local superabundance, and have a diurnal activity pattern (Reynolds and Braithwaite, 2001). Black bears are not predictable in activity or location, are not approachable, are not readily viewable (particularly in the many thickly forested areas on the AT), are generally not tolerant of human intrusion (although Shenandoah-based bears are believed to be some of the exceptions), are not particularly rare or superabundant, are crepuscular, and will choose whether to be diurnal or nocturnal based on which pattern will offer the most foraging success for their individual circumstances. From this perspective, black bears are far from the 'ideal' wildlife-viewing candidate. Indeed, bears are part of what Chris Philo labels the "exclusionary extreme"; animals that we normally keep at as much physical distance from ourselves as possible (1998: 66). It is this, in part, that makes them 'authentic' in the context of the wilderness setting that AT hikers were looking for. In a way, an AT thru-hike is a performance of the "exclusionary extreme", and entering a space that is inhabited by black bears is one indicator of the authenticity of that experience.

Risk

Encountering a bear on the AT can be seen as more authentically wild than organised wildlife tours, not only because of the characteristics of bears listed above, but also because the encounter, like all animal encounters on the trail, is unmediated by any authority figure. Although there is plenty of advice available online from organisations such as the NPS and the Appalachian Trail Conservancy, hikers are essentially on their own when out on the AT; it is up to them how to respond to an animal encounter, how to behave towards the animal, whether to try to engage with the animal or whether to turn and run in the other direction. Aside from perhaps their fellow hikers, there is nobody watching the hiker in their interaction with a bear, and nobody to step in and rescue the hiker should the encounter turn dangerous. For this reason, the element of risk is very real to many of the bloggers who talked about what it might be like to encounter a bear. Bloggers talked about ways of minimising the risks associated with hiking through bear country (for example, using bear bags or canisters to store food, "a more effective means of keeping human food away from any wildlife"), but to eliminate the risk altogether would be to remove an important element from the experience. For many hikers on the AT, part of their motivation in getting 'back to nature' was to remove themselves from modern day society, a common theme in the Trek blog posts, as well as evidenced in Fondren's study (2016). As Patrick Laviolette puts it: "modern societies are obsessed with controlling risk" (2007: 1). Conversely, the wilderness is seen by many as an inherently risky place to dwell, with few elements as risky as the chance of encountering a bear. Laviolette discusses risk-taking sports (such as abseiling, cliff-jumping, white-water canoeing) as ways of reintroducing and then confronting risk, activities that may be viewed as "mild forms of anarchism" (2007: 2). Stephen Lyng (1990) describes "edgework" as the practice of engaging in high-risk behaviours, such as risk-taking sports, as well as other forms of voluntary risk taking, such as selecting a high-risk profession, for example firefighting, wartime combat or even business entrepreneurship. He says that "activities that can be subsumed under the edgework concept have one central feature in common: they all involve a clearly observable threat to one's physical or mental well-being or one's sense of an ordered existence" (Lyng 1990: 857). Although Lyng includes certain types of professions in the practice of edgework, he, like Laviolette, links the imperative towards risk taking to the constraints put upon people by modern life, particularly the deindividualising effect of living and working in industrial and post-industrial societies, and the associated feeling of lacking control over one's life.

Bloggers who anticipated the dangers of living among bears seemed to relish the surprise of their friends and family members that they would put themselves in such a risky situation, with one writing "C'mon, you're talking to someone who has had a bear inside a lean-to with them and lived unscathed to tell the tale!...No I'm not afraid of bears". These types of narratives can be viewed as positioning the author's loved ones in the role of society at large, while constructing themselves as risk takers, edgeworkers, or mild anarchists.

Compromised wildness

Julie Kalil Schutten, in her study of Werner Herzog's film *Grizzly Man*, based on the life of Timothy Treadwell – who chose to live among the grizzly bears of the Alaskan Peninsula, and was ultimately killed and partly eaten by a bear – remarks on the dissonance felt by viewers of the film on realising that a human had become "pieces of meat" for an animal, effectively forcing humans to the nature side of the socially constructed nature/culture binary (2008). In many, admittedly subtler, ways, hiker-bloggers voluntarily moved themselves, or flirted with moving themselves, to the nature side of this assumed binary; they lived outdoors, had no possessions but the ones they carried on their backs, drank from streams or other found water sources, stopped washing or grooming themselves, and generally cut themselves off from modern society – although interestingly choosing to retain links through the writing and posting of their blogs online.

Whilst some hikers viewed themselves as moving closer to nature, there was also evidence that they were disturbed by the apparent moving of bears away from 'nature' and towards 'culture' by seeking out and consuming human food. As one blogger wrote, "If animals come in contact with a hiker's food, the 'wild' in 'wilderness' is compromised", the implication being that the bear's *authenticity* itself is compromised, *both as a wild individual and as a standard bearer for the wilderness as a whole.* Ingold (1994: xxii) refers to the common attitude that "the wild animal that lives an authentically natural life is one untainted by human contact", and Cassidy (2007: 8–9) speaks about domesticated animals as having been viewed as "corrupt and inauthentic versions of their wild ancestors". Much like the dissonance that Schutten emphasised as coming from humans seeing humans as meat for the first time, blogger narratives provide indications of a dissonance rising from the awareness that (natural) bears are seeking out (cultural) human food and obtaining it.

Bulbeck (2005) references the parameters for authentic wildness that humans tend to put in place in *Facing the Wild*, her study of ecotourists seeking out close encounters with animal Others.

> [N]ot only is Western philosophy shot through with dualistic ways to carve up our understanding of the world, but the search for 'authenticity', and the spiritual yearning associated with this, may be a peculiar quest for contemporary industrialized Westerners. This desire makes it harder for us to give up the idea of untouched wilderness, of wild free animals
>
> (Bulbeck, 2005: 191)

In expecting bears, some hikers anticipated the excitement of encountering a bear, while at the same time warning each other against the dangers of an encounter not only to themselves, but also to the bear. They anticipated that there was a risk of physical danger to themselves, but for the bear there was both the physical danger of becoming habituated to humans, becoming a 'nuisance' bear and ultimately being killed, and the implied moral danger of being 'tainted' by human contact and no longer being authentically wild. If a bear were to stop being authentically wild then his status as metonym for wildness is compromised. As the presence of bears – the only large predator on the trail – is used as 'proof' that the AT is indeed a wilderness this could also mean that the whole status of the AT as a wilderness trail is called into question. For people wanting to undertake a wilderness adventure it was therefore vital that bears on the trail remain as 'untouched' as possible, in order not to compromise the authenticity of the experience itself.

Living among bears

Not every hiker had a direct encounter with a black bear. In fact, it appears from the blogs reviewed that only a minimum of hikers had interactions with bears while on the trail. Many hikers, however, hiked with a strong sense of the presence of bears along the trail, not only due to prior knowledge, but also because they saw them – sometimes far in the distance and sometimes nearby; they heard them – perhaps during the daytime, but frequently at night; or they were made aware of their presence by others – often from other hikers and sometimes from official notices put up in the areas that the hikers travel through. This awareness of the proximity of bears was construed as thrilling or threatening, or both, often seemingly depending on how vulnerable the hiker felt herself to be. For example, sightings in the distance while walking in the daytime were often reported as exciting and exhilarating, while waking in the middle of the night to the sounds of a bear sniffing around their tent seemed to be the time a hiker felt the most heightened sense of threat or even panic.

For this book 38 blog posts that can be characterised as describing dwelling among bears were analysed. Many hikers gained a heightened awareness of the presence of bears from other people, including one hiker who recounted a story told to her by a ranger, who reported having been hiking in the Smokies and returning to her car to find that the door had been ripped off and a bear was sitting in the back seat, eating an apple. This led the blogger to express her shock at the incorrect practices of other hikers on the trail.

> The bear that broke into her car was able to smell the apple in spite of the closed doors! And yet, I see hikers sleep with their food bags rather than hang them where bears can't get them!

Another blogger described a woman "covered in down", who told her that she and her hiking partner had stopped to set up camp the previous night, and after pitching their tent had gone to hang their bear bags in a tree. While they were gone, a bear had destroyed their tent and one of their sleeping bags and "ran off with their packs". Another hiker "ran into a park ranger while shopping at the Dollar General", who told her that "every year he returns a few dog collars to owners after dogs have run loose and bothered bears. Just the collars, because the remains are usually in bad shape".

Trail registers are notebooks kept at certain points along the AT where hikers can write communications with each other. One blogger described a note left in a trail register by another hiker who was asked by a hunter to help carry the black bear he had killed over to a weigh station. In return for the hiker's help, the hunter gave him a ride back to the trailhead in the back of his pickup truck with the dead bear. Posters were another way that hikers found out about bear presence.

> Bear closure everywhere…I had a plan to walk 12 miles to get past the closed shelter to the next campsite. I reread the bear closure poster and found out that campsite was closed as well so I would have to complete 17 miles…A ridge runner reported that the second campsite that I was headed to was closed as well due to bears stealing food the night before.

Sightings were the most common way that people became aware of their coexistence with bears along the trail. Hikers often posted their 'bear counts'; one blogger was surprised to see five bears in one day so close to winter, another noticed seeing more bears and other wildlife once he started hiking alone, sighting two bears, a turkey, multiple deer, and a few snakes all in one day. Shenandoah National Park, in Virginia, is one of the most well-known places to see bears along the AT, and is renowned among hikers for bears that are supposedly unperturbed by the presence of humans.

> The other notable highlight of the Shenandoahs include ten bear sightings in four days – three cubs and seven mommas and adults. Other than the one bear we heard walking around our campsite, close enough to where we heard him breathing (!!), all the others were safe distances away and none of them caused any trouble. Needless to say, we can go the rest of the trip without seeing another bear and I will be just fine!

Many bloggers were delighted at being able to report their bear sightings, particularly when the bear was engaged in an interesting behaviour, such as the hiker whose friend came across a bear cub raiding a hornet's nest. Another described a bear falling out of a tree near to her.

> By afternoon, the temperature probably creeped up into the 80s, but it wasn't humid, and there was a breeze. Not a bad day for hiking. We just plodded along slowly fighting off the spider webs and the gnats – until a bear fell out of a tree! I didn't know bears could fall from trees, but that's what this looked like. At first it sounded like a huge limb falling from

the sky, but when we looked up, we could see something large and black falling down near a tree about 50 feet in front of us! It was hard to tell exactly what happened, but we backed away, hoping the bear wouldn't blame us for his misfortune!

After waiting for 15 or 20 minutes, we cautiously walked back up the trail yelling "HEY BEAR, HEY BEAR!" (Backfire's call) and "YOOHOO! YOOHOO!" (my call).

That was our excitement for the day!

Another blogger reported feeling so pleased to see a bear that "he motivated me and I ran up the Stairway to Heaven without the use of trekking poles and without even taking a breather", and another wrote "Man, I definitely feel a rush every time I see…a bear!"

A veteran AT hiker wrote that she had seen 13 bears on her thru-hike, as well as many other animals, and posted tips on how other hikers could maximise their chances of seeing more wildlife, which included hiking in the hours around dawn and dusk, putting rubber tips on hiking poles to make them silent against rock surfaces, hiking alone, and tuning "into nature's rhythms".

Yet as well as being conceived of as thrilling and inspiring, bears were also viewed as a threat to hiker safety. In particular, hikers seemed to know well that mother bears with cubs were dangerous to approach, and that a female bear may attack any human that she perceived as a threat to her cubs. Although only one hiker reported having seen a mother bear with cubs 'charge' another hiker, and then later on themselves been the target of a 'charge' by a mother bear with cubs, many (though not all) hiker-bloggers reported an added level of caution taken when aware of the presence of a female bear with cubs, as in the next two accounts:

> BB and I saw seven more bears over the next day, two families and a lone bear. We boogied out whenever we saw a mother.
>
> We left the lake quickly, moving at a fast 3mph pace in order to get away from the lake and through the trail, four miles of which was closed to all except thru-hikers due to a family of bears that had taken up residence in the area (and really made their presence known).

When cubs were seen on their own, hiker-bloggers were often particularly wary of the possibility that a mother was not far away, even when excited about the sighting.

> I was in the zone, hiking fast around a corner, when I looked down and saw a black furry thing. It was a foot away from my own foot, if even that, and for a moment I froze as the cute, furry thing turned around. "Holy shit", I told Seamster, "that's a baby bear!" Very quickly my thoughts shifted. "Holy shit" I blurted out "Where's the momma bear?" Quickly

we got far enough away to avoid a mauling, but close enough to warn the next hikers coming down the trail.

Lone bears were also often perceived as a threat. One hiker mentioned seeing a very large bear close to her and her companion's intended campsite, prompting them to hike further than they had planned, to a more distant campsite. Another admitted feeling nervous when he spotted two bears on two separate occasions along the side of the trail, saying that he felt "more inclined to move along than to pull out my camera". A few bloggers mentioned feeling wary after seeing signs up about a bear presence in the area ("our rainy hike took us through an annoying four mile section where walking was fine, but if you stopped for more than a few minutes a bear would come by and disembowel you. That was the vibe I got from the signs, anyway").

Several bloggers mentioned having heard a bear, or what they suspected was a bear, sniffing around a campsite at night, either outside the site perimeter, or all the way up to tents.

> The Fonz and I saw a mama bear and cub on the way to the shelter. Later that night I was awoken by a bear outside my tent which kept me awake for a while when it decided to hang out. Those bears have been real jerks lately.
>
> There has been a lot of bear activity on the AT this year and the area we were in had a bear activity warning. That night around 4am Patrick and I woke up to a noise in the woods. It sounded like something large was pacing back and forth somewhere close by. We concluded it was a bear after comparing notes with those who stayed in the shelter the next morning...

One blogger wrote that when she and her friend heard the sound of what they thought was a bear moving around their tent in the middle of the night, she barked like a dog, as they had been told previously that bears in the Smokies are hunted with dogs, and barking would scare a bear away.

It was clear from hiker-blogger posts that bears in many places along the AT have become habituated to human beings. American Black Bears are commonly described as shy animals who are naturally frightened of humans and will tend to avoid being in their presence. Yet their habituation to hikers is evident, even when no direct interaction has taken place, as many bloggers spoke about their surprise at a bear's reaction – or lack of reaction – to their presence. Some were delighted at the close proximity that this allowed. One hiker, clearly pleased, stated that

> Bears are so tame and actually hop on the trail and walk it for a little while". Another was happy to have been able to get close to a mother bear and her cubs, and also described bears on the trail as "tame.

Everyone told us that we would see a bear somewhere in the park if we hadn't already in our trip, but our bear count sat steady at zero. That is, until this afternoon at the edge of the park where we got a close encounter with a momma bear and her two adorable bear cubs! They were so sweet and not bothered by us at all – about as tame as the deer we've seen and hardly skittish. Watching them was awe inspiring and the highlight of my day for sure.

Yet the evident habituation of bears to hikers was not appreciated by all, with one person writing that "the bears didn't bother me, but they also didn't run away as quickly as I'd hoped". In particular, the bears' fearlessness regarding proximity to humans was particularly unappreciated when it came to bears entering what hikers considered as their own space: the campsite. Among the many reporting midnight campsite visits by bears, one blogger wrote about a friend of hers, who, when camping on his own, heard more than one bear sniffing around his tent. He blew his bear whistle several times, but it seemingly made no impact on the bears, who remained for a while. The hiker "made it through the night, but he was quite ready to sleep with a group in a shelter the rest of his time in the Smokies".

Dwelling and habituation

In his study on mediated tourist encounters between humans and alligators in Louisiana, Adam Keul talks about "our motivation to experience the spaces to which other animals belong" (2013: 938). Unlike most ecotourism activities, hiking the AT is more about shared space, and the awareness of shared space, than prolonged observation or direct interaction with wild animals. Hikers and bears share the spaces on and around the AT, most often without any direct interaction with each other, but frequently with an awareness of the other's presence. Once out on the trail hikers became conscious of the presence of bears by seeing, hearing, and hearing *about* them. This presence can be perceived as threatening and even frightening, and at the same time be a reassuring reminder of the authenticity of the wilderness experience being had.

Much of a hiker's perception can be attributed to the proximity of the bear or bears to the hiker. In wildlife tourism close proximity between tourist and wildlife has often been identified as a key feature of the experience (see Curtin, 2005; Duffus, 1988; Schänzel and McIntosh, 2010; although see Orams, 2000 for alternate findings); however, for AT hikers the ideal proximity was a middle distance, not too close as to pose an immediate threat, and not too far away that the bear could not be observed relatively clearly. Hearing the bear without seeing him was most often perceived as threatening, particularly when at close proximity. Hearing about the presence of bears in the area from someone else was often interpreted as thrilling, yet necessitating cautiousness.

Dwelling

In *The Perception of the Environment* (2011), Tim Ingold discusses what he terms the "dwelling perspective", which "treats the immersion of the organism-person in an environment or lifeworld as an inescapable condition of existence" (2011: 153). Rather than supposing that we inhabit a world to which form and meaning have already been attached, the dwelling perspective sees the environment as continually coming into being around its inhabitants, and in which "its manifold constituents take on significance through their incorporation into a regular pattern of life activity" (2011: 153). Among the many ways in which an AT thru-hike is different to wildlife tourism or an eco-tourism trip is the hiker's sustained length of exposure to the environment, and consequent immersion in it. Walking the trail becomes the hiker's regular pattern of activity. For those who spend months on the trail, the environment around them becomes their lifeworld. The "form and meaning" that may have been attached to the trail and its indigenous inhabitants before the hiker actually set foot on the trail – generically wild, generically wilderness – can be replaced by a more complex and nuanced appreciation of living in the space of other autonomous beings.

> To know one's whereabouts is…to be able to connect one's latest movements to narratives of journeys previously made, by oneself and others. In wayfinding, people do not traverse the surface of a world whose layout is fixed in advance…Rather, they 'feel their way' *through* a world that is itself in motion, continually coming into being through the combined action of human and non-human agencies.
>
> (Ingold, 2011:155)

Aside from shedding some light on why hikers might feel compelled to blog about their journey along the trail, there is no doubt that hikers who were aware of sharing the environment around the trail with bears were also aware of the bears' status as agentive beings. Although wildlife managers such as the NPS are present around the trail, hikers and bears are largely left to their own devices. Indeed, it is the hikers who are sometimes expected to change their own behaviours, including where on the trail they hike or camp, to make way for a bear's own decisions about where she goes and what she does. In this respect, bears on the AT can indeed be viewed as embodying wildness, in its original meaning of 'self-willed'. Bears and hikers often dwell together around the trail, negotiating the space in a performance that usually – although not always – ensures that each is given enough room to dwell in their own way.

Through many days, weeks, and sometimes months on the trail, hiker-bloggers experienced the presence – and occasionally lack of presence – of bears in different ways. Frequent or numerous sightings of bears would often result in an almost blasé 'old news' attitude to bears. As bears may become habituated to hikers, so hikers also became habituated to bears.

This happened when bloggers mentioned having seen many bears over the course of their hike, or sometimes over the course of just one day. As one blogger mentioned being "over" seeing bears, multiple sightings (or hearings, or hearing abouts) would be reported, followed by no more blogs about bears. It is very possible that more bears were sighted, but it seems that the hikers no longer felt that they needed to be mentioned in their blog, unless the bear was viewed as doing something particularly interesting, or the sighting became a face-to-face interaction. Bears became incorporated into the hiker's regular pattern of life activity, and so part of the hiker's lifeworld, rather than a spectacle to be blogged about. Their presence was proof enough that the hiker was living in 'authentic' wilderness. Once their presence was established, therefore, there was no real need to recount particular incidents. In this way, it could be said that the dwelling perspective encourages habituation to the Other.

Caged bears on the trail

Once a hiker had dwelt with bears in this way, they could experience dismay when faced with the incongruity of seeing black bears in a pen at the zoo that the AT runs through in New York State. One blogger wrote, "So the AT goes through a zoo and the lowest point is across from the black bear cage. So sad to see them pinned in like that when I have seen some in the wild", managing to express apparently genuine disappointment while simultaneously accumulating cultural capital for having lived among bears in the wild. A few bloggers mentioned that seeing the bears in the zoo had been the *only* place that they had seen bears along the entire AT. However, for those having experienced bears as agentive, self-willed beings, going about their business, seeing them displayed behind a fence could be troubling. Jennifer Wolch (1998: 128) cites Alexander Wilson as demonstrating how "urban simulacra such as zoos and wildlife parks have increasingly mediated human experience of animal life. Real live animals can actually come to be seen as less than authentic since the terms of authenticity have been so thoroughly redefined". The bears in the zoo have had their capacity for agency largely removed from them and their wildness/self-willed-ness compromised. Aside from the hiker who mentioned feeling sad at the sight of the bears "pinned in", nobody who said that they had seen the bears in the zoo mentioned having had any emotional response to them; certainly no thrilling, threatening or awe-inspiring feelings were mentioned as they had been elsewhere on the trail. Another blogger wrote that "Black bears sure don't sit around like this in the wild". Only a handful of bloggers mentioned the zoo, or the zoo bears, at all; possibly a combination of genuine perplexity or disappointment at seeing bears trapped in a pen, and cognitive dissonance at seeing a now familiar (through sightings, hearings or hearing abouts) animal so drained of all his 'meaning'. Both the real-life agency *and* the symbolic wildness of bears elsewhere on the trail are missing from the zoo bears. While the caged bear may feel the effects of having had his freedom and agency removed from him, the hiker witnessing the bear in the zoo has had the element of risk removed for her, transforming her from edgeworker to passive observer, and thereby removing some of her own agency as well.

Encountering bears

Out of 166 bloggers, around a fifth wrote about a specific interaction with a black bear (with several bloggers posting more than one story about having encountered a bear or bears). By this count roughly 20 percent of all hiker-bloggers had at least one interaction with a black bear (not counting any individuals who had interactions but didn't write about them). The term 'interaction' here describes an encounter during which both parties acknowledged the presence of the other, and responded to the other, however briefly. Two of the recorded interactions were third-person narratives of encounters that had happened between the blogger's acquaintance and a bear, and the rest were written by the hiker-blogger who had experienced the interaction.

Around two-thirds of all encounter blogs referenced incidents appearing to have happened when people came across a bear unexpectedly (i.e. both humans and bears were taken by surprise), in relatively close proximity. This is probably due to much of the AT forming what is commonly referred to as 'the green tunnel', i.e. it runs through densely wooded forest, thereby making it hard to see far ahead or through the trees, and increasing the likelihood of unexpected encounters.

In just over half of the incidents describing hikers and bears taking each other by surprise, bloggers described the bear or bears encountered immediately running or walking away from them. One blogger said, "I have seen exactly three bears on the AT in the many years I have hiked on the trail. All three times, I only caught a glimpse of their butt as they ran off into the woods". Another wrote that he saw two bears and "both ran like the devil were chasing them when they saw me", and another that "as soon as it saw us it immediately ran down the hill and out of sight". For people who encountered bears that ran away, the experience was often interpreted as thrilling and rewarding, and frequently framed as humorous, as the following three extracts from narratives show.

> THIS IS IT. We saw a bear. It was in – wait for it – Bearwallow Gap. Little buddy was so scared we couldn't help but feel a little bad for him. We heard and saw him scrambling away from us continually looking over his shoulder, which only caused him to stumble then get caught in branches. He was terrified, and that allowed me to enjoy the experience of seeing a bear in the wild at close distance.
>
> We also had our first bear encounter...We were crossing a field of waist high grass, and I saw this rustling along the trail about 100 feet or so in front of me. I stopped. The rustling stopped. A bear head pops up

over the grass. I looked at it. It looked at me. Then it turned around and hustled away. No time for pictures, but a definite adrenaline rush.

Not long after…I turned a corner and gasped, and as I gasped a full grown black bear walking down the trail looked me in the eyes and leaped into the brush. Finally, I'd seen a bear! …I also realised my daydreams about fighting off a black bear with my walking stick were highly inaccurate.

Another hiker described encountering seven bears in the space of 15 minutes, including a mother bear and her cubs, who "walked into the woods as soon as they saw us" and another bear with cubs, who "as soon as they caught wind of us…[they] turned and left". For this individual, the high number of bears encountered so quickly seemed to lessen the thrill, and he concluded, "I think I'm 'over' seeing bears". For another, even watching bears run away was a profound experience, as he wrote, "As they lumbered away and my brain shifted into gear I was overcome by just how thankful I was to be there and see them. They were beautiful creatures…". Another said, "Though we felt blessed to have seen a bear we figured we could check that box and there was no need to see another one!"

Five hikers described specific occasions when they surprised a bear who did not immediately run or move away from them – although others made more general mention of this kind of reaction, for example the blogger who wrote that one of his least favourite things on the trail was "dealing with bears that weren't afraid of people". Not everyone felt threatened by a bear's lack of response to them, as one blogger spoke of two separate encounters in the Shenandoahs: "The bears were both near large camping areas and appeared to be very comfortable with humans…both seemed very neutral about our presence in their woods". In most instances like this, hikers described either leaving the area themselves or attempting to scare the bear away. One blogger described coming across a bear sitting beside the trail about 20 feet ahead of her group. She and her hiking companions waved their hiking poles in the air, spoke loudly, and made noise to try to get the bear to leave, and "the bear panted and looked bored with us and finally crossed the trail to head up the mountain". Another hiker described her dog chasing a bear into the woods, and returning a few minutes later, to her relief.

Of all blog posts describing a direct interaction between a hiker and a black bear, around a third of narratives described situations in which the hiker was deliberately approached, or appeared to have been deliberately approached, by a bear. The approaches usually happened when hikers had either paused for a rest, or had stopped at a campsite for the night. However, one hiker wrote about being followed while on the trail.

Albus heard the sound in the woods and I did not. A few minutes later I did – and it sounded big. Maybe a deer, but it was a heavier, slower, bigger sound. In a bit there was a clearing through the trees and there was our bear. All 400 pounds of him. For the past several minutes he had been slowly pacing us – thirty feet into the woods and about twenty feet behind us. Much scarier than a still bear. Albus kept his cool and we played the same game we had with the earlier bears: high poles and loud, low voices. We kept walking at our same pace and shortly thereafter the bear lost interest in us...We were a little nervous about tenting alone. We hadn't seen other hikers since early in the morning. We braved the night alone with our poles and knives kept close.

There appears to be much evidence of bear habituation to hikers, as some accounts describe bears that were very difficult to scare away. Most of these descriptions centred around an incident at a campsite. This entailed a different situation than bears surprised on the trail, as bears around campsites appear to have deliberately approached them precisely because of the presence of humans, and more pertinently, human food.

Bears in several places along the AT have learned to associate humans with food, and some bears are known to have overcome their natural shyness around humans specifically in order to attempt to gain access to the food that they carry with them (see Clark et al., 2002; Gore et al., 2006; Tate and Pelton, 1983). In particular, this has resulted in accounts by bloggers of night-time 'raids' on their camp sites, although many of these did not develop into a direct encounter, as detailed previously. However, there were some accounts of bears approaching campsites despite the humans present attempting to scare them away. A few hikers described interactions happening during daylight hours.

> [W]e were all getting chummy after I cooked my spaghetti when GP pointed to a log fifteen yards behind me and said "bear". I thought he was joking and only paid mind when he was insistent. Sure enough this medium-sized black bear with a curled lip looked at us as if we were a food bank. He sat earnestly like a dog at the dinner table asking us for scraps. GP yelled at him to "git" and directed us to throw rocks, but the bear deftly jumped in a wide circle around Tom Floyd's Wayside (shelter). He was there a while and even greeted GP when he walked down to Ginger Spring.

These accounts usually involved an acknowledgment that the bear had been attracted by the smell of food, as when another blogger talked about a bear approaching the camp after he and his hiking partner had cooked a dinner of pasta and ham. When they had tried to scare him away by yelling and waving their trekking poles, he "was not so easily deterred. With a head tilt reminiscent of a dog curious to know if you had a treat in your hand, he stood his ground, wondering just how threatening we were". In the morning they realised, too, that their chosen campsite was close to a dump, and concluded that must have been why the bear was nearby in the first place.

> Oh yeah – and we saw an aggressive bear that wouldn't leave our campsite. It was there in the morning...and no matter how much we yelled and clapped at it, it just ignored us and stared at the firepit, where day hikers had been throwing out trash for who knows how long.

Definitions of what counted as "aggressive" behaviour varied, with the previous account indicating that the bear's continued presence at the camp was interpreted as aggression, even though he did not enter the main site or react to the rocks being thrown at him. One hiker recounted being particularly nervous about a bear's night-time presence at her campsite due to having her dog with her, but did not interpret his behaviour as aggressive.

> [T]he entire camp was being stalked by a bear that really wasn't interested in going anywhere else. He wasn't aggressive, but he claimed his territory and decided it was the humans that needed to go, not him.
>
> Apparently, the bear was right at my tent and my dog growled. I was too asleep to notice…Then, I saw a glowing face near my tent again. The bear was stalking us. He wasn't acting aggressive, but curious. Our whistles and hollering never phased [sic] him. He just kept stalking and circling the entire camp area and shelter. I made the executive decision to night hike, and get the hell out of there.

In another narrative, a hiker recalled being scared out of his camp during the night. He was camping with two companions, and while he and his friend had stored their food in odour-proof bags, the third hiker had his food stored in a plastic grocery bag. The blogger wrote that he "awoke to a black bear 5 feet from my tent entrance bent over eating the food" belonging to the third hiker. All three ran away from the camp, and when they returned the next morning to clear up their belongings, found that the bear had eaten all of the third hiker's food supply. He said that "the bear was able to open individual candy bar wrappers, peanut butter containers, and tuna packets".

The following blog post, reproduced almost in full, gives an account of an early-morning interaction between a hiker and a bear at a campsite.

> As I returned to the vicinity of the shelter and headed back to my tent, there, sitting on its haunches in my half-open vestibule, was a large male black bear. The bear eyed me with a haughty malice, daring me to approach.
>
> The National Park Service has a list of suggestions in the event of a black bear encounter. I'm not going to say I deliberately ignored these suggestions, but I may have bent a rule or two. I don't think I need to remind you how rude this bear was being.
>
> Rule Number 1 – remain calm and don't run. Okay, this one is easy, I thought. I'd been running into bears all week, and the result was the same each time. Stand your ground, appear tall, and don't freak out, and the bear will eventually trundle off in search of less prideful prey. However, my calm at seeing a bear move into my home was matched – perhaps even outmatched – by this bear's own implacability. He pawed at the ground a bit, circled back and forth as if he was a dog finding the best spot in its bed, and settled back into its vigilant posture.
>
> "Sorry pal", the bear seemed to say, "but I live here now".

I crossed my arms and stared what I thought was a menacing and alpha stare at the interloper. The bear just made a curt chuffing noise and eased his behind closer to the door of my tent. Okay, Rule Number 1 unsuccessful. What's next?

Rule Number 2 – Let the bear know you are a human. Talk to it in your normal voice and wave your arms.

Alright, this one I know I could do. I am fairly confident that I am a human, now to prove it to this disrespectful creature. I waved my arms around, with hesitance at first, unsure of how the bear might react, and then more and more frantically. Some other hikers had awakened by now, and were watching my tribulations with vague amusement. Leave it to the AT hiker to be thoroughly unimpressed by my predicament. After all, they've seen it all before. After what seemed to be an eternity of well-intentioned arm waving, I gave up. I'm not sure if bears have eyebrows, but I could be reasonably certain that this bear had raised his in a gesture of disinterested contempt. It was time to up the game.

"Hey bear!" I shouted, voice full of confident machismo, "get on outta here!"…

The bear stared at me insolently …

Rule Number 3 – If the bear does not leave the area, move away slowly.

Now this is a rule that I quite openly ignored. This bear was occupying my sovereign territory and I would be damned if I would cede my tent and all the precious sentimental gear stashed within it to such a brute. I took a few tentative steps from side to side, trying to divine the bear's intentions as I circled closer to my tent. I picked up a few rocks from the ground and tossed them at nearby bushes and trees, making as much noise as possible. Surely the bear will be startled and chastened by my sudden and cunning sonic attack, I thought.

Startled he was, but chastened he was not. He reared up briefly on his hind legs and then drove his front paws into the dirt with an audible slap. Bear biologists and Shenandoah Park Rangers will tell you that this gesture is not as threatening as it may seem, "it is merely a bluff that means the bear feels nervous and apprehensive, but for some reason may be reluctant to leave". I had in fact spoken to a park ranger earlier that week who told me that a bear had to be removed from the park recently, because it had grown clever enough to bluff charge hikers like this and frighten them into dropping their packs and running away, at which point it could rummage through their selection of trail mix and candy bars unmolested. A shrewd creature, indeed…

For my part, I momentarily forgot this nugget of information, when confronted with the stamping feet of the 300 pound behemoth who had usurped my tent, and I was justifiably scared. I backed away a few steps, a signal the bear was sure to interpret as submission and surrender. However, as a last gesture of defiance and strength, I made sure not to break eye contact.

Rule Number 4 – Avoid eye contact with the animal.

Oops. Maybe this was why the bear was being so disrespectful to me! I had been staring him down the whole time, assuming that a stink eye and a breath of animal confidence could frighten him away... .

I figured at this point that the only way to usher this bear out of my tent and out of my life forever was a complete sensory assault.... . Lacking a megaphone or a boombox, I slowly retreated backwards to find the next best thing. Keeping an eye on my foe the whole time, I groped around on the shelf inside the Byrd's Nest for my eating utensils. Brandishing my titanium spork and stainless steel mug, I advanced on the bear again with my improvised sword and shield at the ready.

CLANG CLANG CLANG CLANG CLANG!

The spork rang out against my coffee cup, filling the still mountain air with reverberations of sound sure to be menacing to the most hard hearted of bears. And indeed it was! My enemy (for whom I had developed a grudging respect by this time) lifted his rump into the air as if my tent had suddenly caught fire and went crashing away into the underbrush. I had successfully liberated my possessions back from mother nature...

Author's note: This is a story of a black bear habituated to humans, the kind that is especially dangerous to hikers. I got lucky this time, but take the advice from the National Park Service seriously, and cultivate your awareness, especially in Shenandoah!

Like many hiker-bloggers describing potentially frightening bear encounters, this blogger uses humour, and his narrative has been styled for dramatic and humorous effect, framing the whole interaction as a comical battle of wills. Another blogger wrote about being approached by a bear at a campsite that was "within the supposedly safe zone". Her dog's barking seemed to scare the bear away.

As noted previously, during 2016 several bloggers wrote about how shelters in the Smoky Mountains National Park had to be closed due to bear activity in the area, including bears approaching and entering shelters while hikers were still in them. One blogger described having heard that due to a low acorn crop the previous year bears didn't have as much food as they needed, and "with all the hungry bears there is a bit of chaos throughout the shelters in the park". The same blogger went on to describe an incident in which one of the people she was hiking with left his bag on the ground to go to the privy, and when he came out he saw a bear dragging it up the mountain: "you could hear the bear ripping apart his bag to get to the food". The group of hikers waited until the bear was finished with the bag before they went and retrieved it. Later on, after arriving at their shelter, the hikers were approached by rangers from the NPS, who set traps around the shelter. The following day a bear set off one of the traps, and the hikers were able to watch as the rangers tranquillised the bear and took blood and fur samples. The hiker did not say what happened to the bear after the samples were taken, but she said, "we learned so much about black bears from the rangers while they did their work". Another blogger described arriving at Spence Field Shelter a few days after a bear attack – probably the one on Bradley Veeder (Veeder, 2016), and found that park rangers were building an electric barrier around the shelter. Later, while walking up the trail, the blogger describes coming across a large bear.

> I got a picture with my camera, but I'll have to wait to upload it until I get home. Turns out he was probably the one who made the attack: the rangers tranquillized him only a few minutes after I went by, and made the ID, so I took the last picture ever of that bear...at the time of this post the DNA hasn't been confirmed so I am editing out some of the information for later. He took a good look at me and then turned away: he was clearly not concerned about me at all. He looked like a beautiful, healthy animal, the rangers said he was about 20 years old and 400 pounds, but I guess his teeth were going bad and that can cause the sort of campsite pillaging that led to the attack. I feel pretty bad for the bear, but it may be for the best. Without good teeth the rest of his life was going to be a hard struggle.

Out of 166 bloggers, only one person described deliberately wanting to find and interact with a bear. When hiking with four companions they came across a sign warning about a 'problem' bear in the area, meaning that the shelter they had been heading towards was closed.

> "So what does this mean?" Scone Boy asked, "should we stop here, or keep going? Cause the sign says we can't camp within five miles of the problem bear shelter".
> "You think there's gonna be a park ranger waiting for us at the shelter?" I asked sarcastically.
> "No, but I'm not trying to fuck with this bear tonight", he replied.
> "Dude, if we want to go to the shelter we probably won't even see the bear, and if we do so what, there's five of us!" I pronounced confidently.
> "If we go there we're just asking for trouble. I've done enough stupid shit in my life, I can't be running around looking for bears," he said, waving his hands passionately.
> "Oh come on dude!" I said excitedly, looking around to everyone else, "Let's go there. Fuck the sign and fuck that bear. We could take him."
> Clovis and Good Talk started laughing.
> "Come on guys it'll be fun!" I continued, "I wanna see this bear".
> ...
> "A head kick wouldn't work on a bear".
> I shrugged. "Probably not, but I'd still throw it. What else am I gonna do? Lay down for him?"
> "No. But you want to see the bear. You're looking for trouble."
> "I'm looking for adventure".

The blogger says, "a life without risk sounds miserably boring to me", but eventually his group set up camp away from the shelter. He described hearing screaming and shouting coming from another camp close by, but he and his companions decided that it was just a loud party, and lay down to sleep. Shortly afterwards, a bear entered their camp, and the narrator describes himself and the rest of his group jumping up to confront the bear, shouting and clicking their trekking poles together.

"Get out of here bear!"
 "Go away bear!"
 "Fuck off bear!"
 Nice Chat started roaring at the bear, channelling his most primal self. I found that to be amusing, but I held it together, trying to be as serious as possible about the situation at hand. The bear began to back up slowly, and the human wall we formed began to close in on him, however hesitantly. I felt braver and braver the louder I yelled, and my confidence grew like a raging fire inside me. It culminated in me thinking, I can take this motherfucker. I really believed it. If this bear comes at me I will fuck him up.
 "I think I can take him", I said to everyone's surprise. Then I stepped forward boldly, out of formation, and yelled, "Get out of here!" He looked up, directly into my eyes, and I stared right back, showing my teeth like some kind of savage. When two animals are about to fight they make direct eye contact. This bear might actually charge me. Then I realised I made a grave mistake. What am I doing? I don't want to fight this bear. Clarity came to me, and I broke eye contact, looking down for a moment, and then stepping back into formation.

The narrative describes the bear backing away and leaving the camp, then returning again a while later.

We were far more confident this time around, and we were mentally prepared. We formed our human wall again, yelling and roaring at the bear, who already seemed reluctant to get any closer. He backed up behind a tree, peeking out from behind it, half exposing himself.
 "Someone throw something at him," Scone Boy put his idea forward and Indiana replied, "Got it".
 Picking up a stick, he aimed and chucked it at the tree like he was throwing a tomahawk. The bear was visibly startled and moved out into the open. We all picked up sticks, still yelling in scattered voices, and began throwing them at the bear.

This time the bear ran away and did not return to the campsite.
 Of 90 blog posts featuring bears, four posts describe a bear actually running toward a hiker, "charging" as hiker-bloggers described it, or bluff charging as

it is sometimes called. In these four posts, three separate charging incidents are described (two posts discuss the same incident). One blogger talked about witnessing a mother bear with three cubs charging another hiker as he "tried to sneak by on the trail, clicking his poles together frantically". A few days later the same blogger was herself "charged" by a mother bear.

> Fauna and I rounded a corner in a section with tall brush on the trail, and I saw a bear cub shoot up into the tree next to me, and said "oh shit, that's a cub". Another slightly larger bear stared at us from the bottom of the tree trunk. Fauna screamed. Mama bear was huge, angry, and five feet away from us on the trail. She stood up on her hind legs, and hit her forepaws on the ground, snorting and moving her head side to side. We retreated back up the trail quickly, and blowing our whistles, advanced again, hoping to scare the bears away. The huge bear got up on her hind legs again, and then ran at me in a charge. Forgetting everything I know about bear safety, I turned and ran away with Fauna in tow…we were badly shaken…and honestly, I don't need to see any more bears.

Another writer reflected on her encounter with a charging bear in two separate blog posts, one written shortly after the incident and one after completion of her hike. She was on her own on the trail when she encountered the bear.

> I tried to scare it off but this one was a bit aggressive and decided to charge me. Even though I thought I was going to get mauled I stood my ground, made myself tall, tried not to make eye contact and started talking to it. Luckily it was a bluff and the bear veered off from me. I backed away slowly banging my trekking poles over my head and cautiously moved on. I'm definitely glad I know my bear safety.

Overall, blog posts discussing hiker–bear encounters covered a range of responses from both hikers and bears. When taken by surprise, most bears would run or walk away from hikers, although some ignored hikers and continued what they were doing, and mothers with cubs most often responded to hiker presence with aggression. On the occasions when bears seem to have approached hikers it was when the hikers were stopped and either eating or sleeping. Some hikers tried to scare bears away, through various techniques, whilst others left the area themselves. Bear responses to hikers who tried to scare them away ranged from running away, walking away, staying but making threatening signals such as pawing the ground aggressively, and bluff charging hikers. Encounters were almost always reported as unpleasant, or at least uneasy on both sides. No hikers reported wanting to prolong an interaction and hiker narratives showed an awareness that when bears approached camps they were attracted by human food, not by humans themselves. Subsequent to bear interactions, however, hikers would often contextualise the encounter as a 'good story'.

Hiker interest in bears as self-willed beings

From blogger narratives about what they expected it to be like encountering bears on the trail, it was clear that bears were seen as a metonym for the wilderness of the AT, and for the concept of 'wildness' itself, and were therefore a tool that people could use to demonstrate their competence in the wild (for example, they could write stories about how best to deal with a bear encounter before ever actually encountering a bear). Dwelling among black bears introduced an element of real danger to both human and bear, which qualified the 'wilderness' experience, but also attuned hikers more personally to the embodied nature of bear presence. However, it is hiker narratives about interactions with bears that come closest to viewing a bear – a specific bear, *this* bear – as an individual. Rather than the bear being simply a signifier, or being uncomplicatedly an animal responding to another animal, people's reactions to bear encounters *combined* ideas about 'bear' as a signifier (wild, predator, charismatic megafauna) with frequently detailed descriptions of how the bear responded to them and what that might say about what he was thinking.

Overall, encounters were unpleasant. It is clear that very brief interactions, during which the bear ran or moved away from the hiker, could feel positive and even thrilling to the hiker – if not to the bear. However, any prolonged contact between hikers and bears was at best uncomfortable for both species, and at worst, stressful and potentially harmful to both bear and human. Wildlife tourists have frequently described their encounters with wild non-human animals as life affirming, transformative or even something akin to an epiphany (see Bulbeck, 2005; Curtin, 2005; Hill et al., 2014; Lorimer, 2015; Schänzel and McIntosh, 2010) but this is largely not the case with AT hikers and black bears. The reason that brief interactions during which the bear ran away were so rewarding for the hiker may be that the encounter combined the thrill of seeing charismatic megafauna with the reassurance that the human was still in a position of dominance; the bear was afraid of the hiker. In contrast, prolonged interactions during which the bear either refused to run away or even moved closer to the humans were a reminder of the bear's own agency, shifting the perceived balance of power, and emphasising the physicality of the bear's body in contrast to the hiker's body. Bears who moved away from hikers "ran like the devil were chasing them", "scrambled away", and "hustled" away. They were referred to as "little buddy" and they "leaped into the brush". Bears who did not run away were repeatedly described in terms of their intimidating physical presence: "all 400 pounds of him", "300 pound behemoth", "huge", and a "brute" who "claimed his territory" or was "stalking us". When bears did not run away the full otherness of their bodies – bigger than human, stronger than human, *wilder* than human – seemed to come into focus.

Fear

During a bear encounter, fear of what the bear might do to them was the strongest motivator for the hiker to be interested in the bear beyond what the bear might already symbolise to them. It was fear that made hikers focus so closely on what the bear was doing and how he seemed to be responding to them, to focus on finding signals that might suggest what the bear was thinking, or what he might do next. In "Embodied Encounters Between Humans and Gators" (2013) Adam Keul describes alligator tourism in Louisiana, saying that "in the cypress swamps of Louisiana's Atchafalaya River Basin, fear is shared between people and gators. Fear paradoxically draws and repels human-alligator contacts" (2013: 930). Keul quotes one of his interviewees as stating that "tourists want to know that an alligator can kill them, but they want to know that they are safe too" (2013: 930). As with the alligator encounters that Keul looked at, the predominant emotion shared by bears and humans was fear. When bears ran away the hiker was able to describe the encounter in terms of the fear felt by the bear; the bear was "so scared we couldn't help but feel a little bad for him...He was terrified". Keul described alligator tourists as knowing "that such encounters can only be experienced in a regulated context" (2013: 930), yet hikers and bears found themselves having to navigate unmediated encounters with each other. When the bear exhibited no fear of humans, it was the hikers who became nervous. Tellingly, one writer wrote that one of his least favourite things on the trail was "dealing with bears that weren't afraid of people". To one blogger, a pacing bear was "much scarier than a still bear", and "we braved the night with our poles and knives kept close". Another defended his reaction to a bear that wouldn't leave: "I was justifiably scared". One hiker who was bluff charged described her hiking partner screaming, as she herself "turned and ran away".

Embodied interaction

Because hikers were unable to read facial expressions in bears (only one blogger mentioned facial features: "I'm not sure if bears have eyebrows, but I could be reasonably certain that this bear had raised his..."), it was the bear's movement towards or away – or her stillness – that was almost entirely responsible for the hiker's interpretation of what she was thinking. Merleau-Ponty (2014 [1945]) describes communication as being achieved through the reciprocity between gestures that can be read to find the other's intentions:

> Everything happens as if the other person's intention inhabited my body, or as if my intentions inhabited his body. The gesture I witness sketches out the first signs of an intentional object. The object becomes present and is fully understood when the powers of my body adjust to and fit over it. The gesture is in front of me like a question, it indicates to me specific sensible points in the world and invites me to join it there. Communication is accomplished when my behaviour finds in this pathway its own pathway. I confirm the other person and the other person confirms me.
>
> (Merleau-Ponty, 2014 [1945]: 191)

The embodied nature of the bear–hiker interaction – bear bodies moving in response to hiker bodies and vice versa – helped hikers to interpret what the bear was experiencing from the encounter, in an act of egomorphism (see Milton, 2005), as opposed to anthropomorphism. In his introduction to Merleau-Ponty's *Phenomenology of Perception* (2014[BIB-043] [1945]), Landes describes the perception of agency in another being:

> The other person's body is not an object for me; it is a behaviour whose sense I understand from within, virtually, allowing for a certain gestural communication through the sedimentations and possibilities of my own body schema. Moreover, when I perceive behaviour, the world immediately becomes the world intended by this behaviour; it is no longer *my* world alone.
>
> (Landes, 2014: xliv)

Thus, the bear's gestures could be read by the hiker as having a self-determining agent behind them, and the hiker was able to experience themselves as the object of a nonhuman person's intentionality.

In Buller's paper "One Slash of Light, Then Gone" (2012), the author explores the idea of embodied movement, "as a means to experiment with a new mode of connectivity and 'being' with non-human animals" (2012: 140), stating significantly that "in movement lies agency" (2012: 145). Inspired by von Uexkull (2010) and Haraway (2008) he continues:

> Through movement, non-human (and human) animals define not only themselves, but also space and time as their own... It is principally through movement that animals have been seen to have agency in our anthropocentric world. That physical, corporal and motile agency that attracts our attention, whether it be through acts of 'resistance' or the active 'co-construction' and 'co-assembly' of the world through presence and vitality...
>
> (Buller, 2012: 145)

Buller adds that "there is another dimension at work here too; that of being or becoming interesting" (2012: 149). As Milton (2002) and Bird-David (1999), amongst others, have pointed out, "our sensitivity to the personhood of non-human animals depends on the intensity with which they engage our attention and respond to what we do" (Milton, 2002: 50). Because of the potential threat perceived in bear presence, hikers paid close attention to the movement of bears that they encountered; the bear became interesting in a way that bears in the zoo that the AT passes through never could.

Many narratives of encounters showed an implicit acknowledgment of bears who did not just 'react' to the hiker but were perceived as *responding* to them; 'animal instincts' dictate a 'reaction', but when an *individual* responds to a stimulus, a *decision* has been made about which response to choose (symbols don't make decisions!). Ingold (2011) cites James Gibson's *The Ecological Approach to Visual Perception* (1979), in which Gibson notes that "other persons and animals" in the environment of the perceiver have the "peculiar ability to 'act back' or, literally, to *interact* with the perceiver" (Ingold, 2011: 167). Having read about bears, talked about bears, seen pictures of bears, related to bears as metonyms for the wilderness in general, actually interacting with a bear made him thrillingly present, thrillingly – sometimes frighteningly – *real*. His behaviour towards the hiker was analysed, reconstructed, and presented in narrative form, in part for other hiker-bloggers to compare to their own experiences with bears. Hikers made judgements on what the bear was thinking at the time.

> [E]ven if [a] tourist's understandings of alligator emotions, behaviour, etc. are vulgarly anthropomorphic – with no ethological basis – they are still meaningful and useful insights. Especially, if being together in space or contextualised embodiment is the justification for our ability to know other species, then why privilege experts?
>
> (Keul, 2013: 946)

Bloggers noted how different individuals reacted to them in different ways. "The first bear scampered off...the second bear...panted and looked bored with us and finally crossed the trail to head up the mountain...yet another bear...had been slowly pacing us – thirty feet into the woods and twenty feet behind us". Thus, rather than being a collective, bears became individuals, and the question for some then became *how will this particular bear respond to me?*

Were hikers interesting to bears? Some bears appear to have ignored hiker presence, others responded to it, seemingly either because the hiker was a potential threat, or because the hiker was a potential source of food. Yet there is also the possibility that bears found hikers interesting because hikers were doing interesting things. Hikers described themselves yelling and clapping at bears, waving their arms around, clicking their hiking poles over their heads, throwing rocks, blowing whistles, standing and staring, or turning and running away. Evernden (1992: 107–108) criticises the view that wild animals couldn't possibly be interested in us, arguing that it removes all subjectivity from the animal. However, Evernden (1992: 118) and Schutten (2008: 202) both cite the naturalist John Fowles's assertion that "All nature...is anti- or ultra-human, outside, and has no concern with man". Schutten (2008: 202) says that "Working toward a dialogue with 'nature' requires that two entities open themselves up to being influenced by each other". Blogger narratives indicated that, whether or not each found the other interesting, neither hikers nor the bears that they sometimes encountered were interested in a dialogic relationship with each other. For bears and hikers, there was no happy interaction because hiker and bear could not both get what they wanted. Hikers wanted to be able to move through the terrain freely, and for those who wanted to 'experience' the wildlife on the trail, they generally (although not always) preferred to be able to see bears from a safe distance. Only one blogger talked about actively wanting to seek out an interaction with a bear. Bloggers also attributed any approach by a bear towards them or their camp to the bear's desire to seek out human food, not human company. As one wrote: "that bear has changed its behaviour because it has learned that it can find food at this campsite".

Telling stories about bears

Where much human–wildlife conflict happens between local people and local animals (see Hurn, 2015 on baboon–human conflict in Cape Town; Rohini et al., 2016 on human–elephant conflict in Southern India; Senthilkumar et al., 2016 on human–wildlife conflict in Tamil Nadu), the animals on the AT are locals but the people are not (despite one of the bloggers claiming wryly that "this bear was occupying my sovereign territory"). Living on the AT for up to seven months, AT thru-hikers were not quite locals and not quite tourists, but resided somewhere in between, as travelling inhabitants. Like people who live alongside autonomous animals, hikers lived with the threat of damage to their possessions or even to their bodies (see Veeder, 2016). Yet, like tourists, hikers were looking for interesting and 'authentic' things to happen to them: experiences that would make good stories (see Urry, 1990). There are many other dangerous animals on the AT: snakes, moose, bobcats, spiders, ticks (who carry Lyme disease), giardia parasites, and hunters (hunting is allowed on at least some part of the trail in every state that it passes through and people have been shot before when mistaken for deer). There are also the non-animal risks of hypothermia, frostbite, heatstroke, falling and breaking a limb, and several other possible eventualities. Yet a narrative involving a close encounter with a bear trumps almost any other incident in terms of cultural capital for an AT hiker. This is probably because being charged by a bear is at once the quintessential experience that people talk about happening on the trail, and at the same time actually relatively unlikely to happen. The importance of risk to many of the people who choose to hike the AT has already been addressed; the component of risk adds more of a feeling of authenticity to the experience (see Laviolette, 2007; Lyng, 1990). Indeed, Reynolds and Braithwaite (2001) write about how the possibility of harm to the tourist can actually add considerably to a person's experience. Ultimately, bear–human conflict narratives make good stories because of the possibility that the hiker could have been harmed during the encounter, which in turn is proof that the hiker has experienced 'authentic wilderness'. As there was no mediator to an authentically wild experience like this, it was up to the hiker to navigate through the encounter without coming to any harm. Hiker-bloggers then translated the fear and adrenaline that they felt during the interaction into a narrative that formed part of the grand narrative of their whole wilderness adventure on the AT, thereby assimilating themselves into the idea of 'wildness', and contributing to the image of themselves as risk takers or "edgeworkers" (Lyng, 1990): "It was a bluff charge. The bear blocked the path behind me so I continued ahead, arriving at the buffet right on time to meet my friends for dinner with a decent survival story in hand"

Furthermore, as Bourdieu puts it, "...the best measure of cultural capital is undoubtedly the amount of time devoted to acquiring it" (1986: 54). The chances of wandering onto the AT for the first time and coming across a bear are slim; encountering one means that you have probably been out in the wilderness for a substantial amount of time, and are therefore more experienced than most and have more authority to speak about life in the wilderness.

Yet telling stories is not simply a way of accumulating cultural capital, or of painting an aspirational picture of oneself. Crafting a narrative can also be a way to make sense of events, particularly in an unfamiliar and often bewildering world. Hiker-bloggers chose different ways in which to contextualise their bear encounter narratives, most frequently framing them as thrilling, threatening or humorous, largely through their retelling of their own response to the bear. Those that were written as thrilling or exciting incidents tended to be about a short bear encounter that ended in the bear running or moving away from the hiker, and therefore involved little or no conflict, including little action on the part of the hiker, but "a definite adrenaline rush". Most blog posts about bear encounters were framed as threatening to a greater or lesser degree, with the ones involving the most conflict (for example, a bluff charge by the bear) usually also written as the most frightening: "The huge bear got up on her hind legs again, and then ran at me in a charge. Forgetting everything I know about bear safety, I turned and ran away...". Yet several bloggers used humour in their narratives, even when writing about situations that were clearly dangerous and must have been frightening, perhaps most notably the hiker who wrote about a stand-off with a bear sitting in his tent vestibule. Framing a frightening situation as comical may seem incongruous but, as Niko Besnier has pointed out, the use of humour has a function in stories that deal with anxiety-inducing situations. Besnier (2016) studied humour in the narratives of the Polynesian inhabitants of Nukulaelae Atoll, at a time when they were trying to adjust to new technologies that had been brought to them by the outside world, specifically a radio-telephone that would allow locals to communicate with friends and family in the country's capital, Funafuti. People who found the new technology difficult to use, and were seen publicly to be having difficulties with it, subsequently told comical stories based around their own anxiety and embarrassment at not knowing how to use the radio-telephone. Besnier spoke about the role of comical self-deprecation in his subject's narratives, describing it as an "equally effective and equally ambiguous form of engagement with a threatening and anxiety-provoking...world" (2016: 90). For Besnier's subjects, it was the idea of having to function in a modern world, with all its new technologies, that was intimidating. For AT hikers, it was the impact of having to survive outside of the modern world, in a world with large, autonomous animals in it, that could be disorienting.

> I waved my arms around, with hesitance at first, unsure of how the bear might react, and then more and more frantically. Some other hikers had awakened by now, and were watching my tribulations with vague amusement. Leave it to the AT hiker to be thoroughly unimpressed by my predicament.

Besnier describes self-deprecating humour as a way of showing humility, "one's own self-abasement in the face of situations that are beyond one's capacity to remain in control" (2016: 83). Furthermore, he points out that narratives in general can function as ways of creating order out of a disorderly past, enabling us to organise the present. Besnier cites Jackson: "to tell a story is to immediately put a distance between oneself and the event with which the story is concerned. A degree of agency is recovered..." (Jackson, 2002: 186). Thus, telling stories about encounters with bears on the AT can be seen not simply as ways of growing cultural capital but, more intimately, as ways of regaining agency out of experiences that may have been unsettling or bewildering – "be-wildering" meaning literally to be "lured into the wilds".

Unlike the Polynesian people that Besnier was writing about, however, AT hikers making sense of their bear encounters by writing stories about them have a pre-existing cannon of stories to relate theirs to; that of the 'wilderness myth', and the 'adventurer meets wild animal' narrative so prevalent in the mythology of wilderness exploration (see Krakauer, 1996; Nash, 1982; Oelschaleger, 1991). Thus, hikers are able to make sense of their bewildering experiences in the wild alongside a history of other adventurers' bewildering wilderness experiences. A blogger talking about his fellow hiker, "roaring at the bear, channelling his most primal self", and throwing a stick "like he was throwing a tomahawk" or his own feelings as he "stared right back, showing my teeth like some kind of savage" is an overt harking back to the 'grand' tradition of wilderness exploration, but the counter-narrative, in which blogger-hikers admit to screaming and running away from a charging bear, is just as relevant, focusing as it does on the undeniably powerful physical presence and fearsomeness of the bear – a worthy opponent being an essential element of wilderness mythology.

Ultimately, although hikers developed a deeper understanding of bears as individual beings during their uncomfortable encounters with each other, it was through the re-telling of these encounters that bears were once again employed as signifiers of wilderness, occupying a role that veered between protagonist and prop in the narrator's account of their wilderness experience. Thus, as with so many meetings between hikers and nonhumans on the AT, there was a continual tension between hikers seeing bears *as themselves*, and as seeing them as *meaning something*.

Bear necessities: the social creation of bears by hikers

> I can never completely coincide with the pure thought that constitutes even a simple idea; my clear and distinct thought always makes use of thoughts previously formed by myself or others, and relies upon my memory…or upon the memory of the community of thinkers…
>
> (Merleau-Ponty, 2014 [1945]: 42)

Merleau-Ponty argued that our perception of something is always inevitably informed by "thoughts previously formed" by ourselves or others, in other words, the pre-existing cultural construction of the entity perceived. Throughout hiker-blogger narratives, inherited mythologies about bears collide with lived experiences of bears, resulting in a complex picture of 'what a bear is like'. Bears have here been shown to be considered as metonyms for wildness/wilderness, but hikers used many other identity constructions, albeit less frequently, in talking about bears. The most common of these can be described as Beast, Pet, Innocent, Deviant, and Science Project. Before concluding this chapter, it is worth looking briefly at these alternate – and contradictory – bear identities.

The social construction of other-than-human animals as a group has been looked at in some considerable depth and is multi-faceted to say the least (see Arluke and Sanders, 1996; Evernden, 1992; Ingold, 1994). When looking at the construction of specific species it becomes clear that there is rarely a single way of thinking about a particular animal. Rats can be constructed differently depending on whether they are seen as plague carriers, food, tools for sport, tools for lab experiments or "a sign of unhygienic conditions and poverty in general" (Edelman, 2002: 3). The consistency in all of these notions is the idea of rats as 'expendable'. Russell (1995) found in her study of ecotourist constructions of orangutans that her informants had composed two dominant stories about them: Orangutan as Child and Orangutan as Pristine. Both stories can be seen as fairly congruent with each other. However, Gore et al. (2011) found in their study of shark diving websites that sharks were alternately viewed as Victims and Perpetrators, in a binary opposition similar to those used by hikers for bears. It could perhaps be argued that the predator status of both shark and bear introduces an added element of complexity to the way that we as humans think about them.

Beasts and pets

The notion of the Beast, and its association with baseness and untameable danger, can be seen as inherently opposite to Pet, with all its connotations of loveable domesticity. Beasts inspire fear, pets inspire affectionate condescension. Hiker references to bears as Beasts are perhaps the most historically determined of the main identities constructed (see Quammen, 2003), harking back to notions of man-eating predators lurking inside an untamed wilderness. Hikers spoke of a "giant, toothsome beast", "300 pound behemoth", and "large, dangerous creatures" capable of "terrorizing" hikers, their language occasionally reminiscent of even earlier mythologies than those summoned by the Europeans first arriving in North America. In *Monster of God* (2003) Quammen describes the lands of Sigurd and Beowulf, in which "there were no native lions during the early centuries of our era…there were no crocodiles or giant snakes in those latitudes. There were no notable lodes of ceratopsians. But there were bears…" (2003: 276), pointing towards an idea of bears having lodged themselves as a threat in the human psyche since ancient times.

Yet just as frequent as Beast imagery in hiker narratives were references to the bear as Pet. Hikers described bears as "Winnie the Pooh", "little buddy", "cute, furry thing", and "adorable". Another hiker wrote, "he's so fluffy I'm gonna die", and another referred to a "round, fuzzy, black bear rump". Dog imagery in particular was used regularly, with one hiker describing a bear settling on the ground as like "a dog finding the best spot on its bed" and another wrote of a bear looking at him with "a head tilt reminiscent of a dog curious to know if you had a treat in your hand". One person even described encountering bears when "for a second my mind filed them as really big dogs, and I had a…brief desire to pet one". This attitude is resonant with comments made by Ladino, remarking on Timothy Treadwell's naming of the grizzly bears that he lived among, that this attitude verged "on a kind of mastery" (2009: 80). In his guide to hiking the AT, Davis describes the bears on the trail as "essentially giant raccoons" (2012: 136), using the image of bears scavenging for human food as their defining characteristic. It could be argued that despite the bear's obvious physical advantage over the hiker, it is difficult to let go of an assumption of dominance.

Innocents, deviants, and science projects

For some hikers, bears were seen as Innocents, in a similar vein to how Russell's informants saw orangutans as Pristine (1995). Hikers described "shy…curious creatures", a "beautiful creature", and "this innocent bear". Yet the image of bear as Deviant was also prevalent. Indeed, the Deviant bear was seen as a once Innocent bear, who through close proximity to humans and their food had become deviant in his behaviour. The Deviant bear, otherwise known as "nuisance bear" throughout several of the studies mentioned previously, was described by hikers as "problem bear", "aggressive bear", "rogue bear", "interloper", and "not the kind of teddy bear I want". Bears who approach hikers for food are widely regarded as transgressive, despite being in "a situation where the parties involved – in this case humans and animals – seemingly cannot even begin to share the same systems of (political) meaning" (Philo, 1998: 52). Hurn (2012: 79) points out that despite human–animal conflict often coming about when humans encroach into the natural habitats of nonhuman animals, it is the animals who are considered as transgressive, rather than human encroachment being viewed as problematic.

A transgressive bear is regarded as having to be 'managed' in some way, which leads on to the notion of bear as Science Project. Interestingly there were no references to this in 2015, yet frequent mentions during 2016, a year also marked by high levels of bear activity on and around the trail, and bloggers posting narratives about NPS rangers carrying out bear management strategies around shelters and campsites.

> Everyone who was still left at the shelter got to watch as the rangers tranquillized the bear, and took blood/fur samples. We learned so much about black bears from the rangers while they did their work.

Hiker-bloggers spoke matter-of-factly about the necessity of bear management, and the ultimate need to 'deal with' bears that were deemed to be too aggressive. A hiker who had encountered a bear near her campsite wrote, "I wanted to make a note in the shelter log, since I understand park service employees check those...but the log seems to be missing. Oh well, the next time I have service I will call it in". The same blogger referred to rangers killing a bear that she had taken the last ever photo of, explaining that "at the time of this post the DNA hasn't been confirmed so I am editing out some of the information for later". Another described attending a lecture during which she learned about the average number of bears per square mile in the area. Evernden (1992: 131) writes that "the wild other disappears the instant it is demystified and saved as a managed resource", while in *The Social Construction of Reality* (1991[1966]), Berger and Luckmann describe the desacralising influence of science:

> Science not only completes the removal of the sacred from the world of everyday life, but removes universe-maintaining knowledge as such from that world. Everyday life becomes bereft of both sacred legitimation and the sort of theoretical intelligibility that would link it with the symbolic universe in its totality. Put more simply, the 'lay' member of society no longer knows how his universe is to be conceptually maintained, although, of course, he still knows who the specialists of universe-maintenance are presumed to be.
>
> (Berger and Luckmann, 1991: 130)

If Berger and Luckmann are correct, the Science Project identity narrative is capable of replacing all other identity constructions around bears – Beast, Pet, Innocent, Deviant, and even Wildness. Yet the reduction of bear identity to Science Project does not imply that myth has been replaced with reality; indeed, the Science Project construction is in some ways as much of a myth as any of the other themes, and arguably does not bring anyone closer to the 'real' bear. It is evident from hiker blogs, however, that bears as Science Project could conceivably replace the prevalent myth of bears as Wilderness in the future, given the intensity with which they are currently being monitored and managed around at least some of the areas of the AT. As Peace says in talking about dingoes in Australia: "they are currently the target of a pan-optic regime which dictates that they exhibit and adhere to certain behaviours and not others" (2002: 19).

Conclusion

This chapter has examined hiker experiences of anticipating, dwelling among, and interacting with black bears, as well as perceptions based around the con-struction of multiple bear 'identities'. The prevailing mythology around black bears on the AT is that they stand in/for Wildness/Wilderness, although this is threatened by the new construction of bear as Science Project. Yet notions of the bear as metonym for Wildness/Wilderness, or as Science Project, though very different constructions, are two types of myth, neither offering access to the 'authentic' bear.

From hiker narratives it is clear that there is a tension between the cul-tural construction and lived experiences of bears. As Merleau-Ponty said, "reflection can never make it the case... that I cease to think with the cultural instruments that were provided by my upbringing, my previous efforts and my history" (2014 [1945]: 62). Yet the embodied experience of living among bears presents at least the possibility of knowing them better. Due to the danger of face-to-face encounters with bears, hiker engagement with them most frequently involved an engagement with the environment they were in as belonging to bears: in "communion ('fellowship') rather than in communi-cation ('conversation')" (Bulbeck, 2005: 149). Hurn writes: "While 'culturally grounded' received wisdom certainly has a part to play, the role of individual experiential engagement with other beings (human and nonhuman) is...an important consideration" (2012: 78). Indeed, there is no reason for hikers to *either* think of bears in terms of the mythology surrounding them, *or* think in terms of what they have directly experienced around them. Arguably the incorporation of all dimensions can lead to a more nuanced appreciation of bears overall.

In *Monster of God* (2003) Quammen interviews a shepherd living and working in a Romanian forest. When asked who is the more troublesome predator, wolf or sheep, the shepherd answers unequivocally, "*Urs*".

> Would it be better, I ask, if there were no bears at all? Well, better for *him*, yes, it would be. But the bear, it's *podoaba padurii*, the treasure of the forest. "If you lose this, you lose the treasure", he says. "A forest without bears – it's empty".
>
> (Quammen, 2003: 287)

References

Adkins, L. M. 2000. *The Appalachian Trail: A Visitor's Companion.* Birmingham: Menasha Ridge Press.

Arluke, A. and Sanders, C. R. 1996. *Regarding Animals.* Philadelphia: Temple University Press.

Baptiste, M. E., Whelan, J. B. and Frary, R. B. 1979. Visitor Perception of Black Bear Problems at Shenandoah National Park. *Wildlife Society Bulletin* 7(1): 25–29.

Beeman, W. O. 1999. Humor. *Journal of Linguistic Anthropology* 9(1/2): 103–106.

Berger, P. and Luckmann, T. 1991[1966]. *The Social Construction of Reality.* London: Penguin Books.

Besnier, N. 2016. Humour and Humility: Narratives of Modernity on Nukulaelae Atoll. *Etnofoor* 28(1): 75–95.

Bird-David, N. 1999. "Animism" Revisited: Personhood, Environment and Relational Epistemology. *Current Anthropology* 40(Supplement, February): 67–91.

Bourdieu, P. 1986. The Forms of Capital. In: J. E. Richardson (ed.) *Handbook of Theory of Research for the Sociology of Education*, pp. 241–258. New York: Greenword Press.

Bulbeck, C. 2005. *Facing the Wild.* London: Earthscan.

Buller, H. 2012. "One Slash of Light, Then Gone": Animals as Movement. *Editions de L'Ehess* 189: 139–153.

Burghardt, G. M., Hietala, R. O. and Pelton, M. R. 1972. Knowledge and Attitudes Concerning Black Bears by Users of the Great Smoky Mountains National Park. *Bears: Their Biology and Management* 2: 255–273.

Carney, D. W. and Vaughan, M. R. 1987. Survival of Introduced Black Bear Cubs in Shenandoah National Park, Virginia. *Bears: Their Biology and Management* 7: 83–85.

Cassidy, R. 2007. Introduction: Domestication Reconsidered. In: R. Cassidy and M. Mullin (eds.) *Where the Wild Things Are Now: Domestication Reconsidered*, pp. 1–25. Oxford: Berg.

Clark, J. E., van Manen, F. T. and Pelton, M. R. 2002. Correlates of Success for On-Site Releases of Nuisance Black Bears in Great Smoky Mountains National Park. *Wildlife Society Bulletin (1973–2006)* 30(1): 104–111.

Clark, J. E., van Manen, F. T. and Pelton, M. R. 2003. Survival of Nuisance American Black Bears Released On-Site in Great Smoky Mountains National Park. *Ursus* 14(2): 210–214.

Clark, J. D, van Manen, F. T. and Pelton, M. R. 2005. Bait Stations, Hard Mast, and Black Bear Population Growth in Great Smoky Mountains National Park. *The Journal of Wildlife Management* 69(4): 1633–1640.

Cole, G. F. 1974. Management Involving Grizzly Bears and Humans in Yellowstone National Park, 1970–73. *BioScience* 24: 335–338.

Curtin, S. 2005. Nature, Wild Animals and Tourism: An Experiential View. *Journal of Ecotourism* 4(1): 1–15.

Davis, Z. 2012. *Appalachian Trials.* UK: Good Badger Publishing.

Duffus, D. A. 1988. Non-consumptive use and management of cetaceans in British Columbia coastal waters. Unpublished PhD dissertation, University of Victoria, BC, Canada.

Eagle, T. C. and Pelton, M. R. 1983. Seasonal Nutrition of Black Bears in the Great Smoky Mountains National Park. *Bears: Their Biology and Management* 5: 94–101.

Edelman, B. 2002. "Rats Are People, Too!": Rat–Human Relations Re-Rated. *Anthropology Today* 18(3): 3–7.

Evernden, N. 1992. *The Social Creation of Nature.* London: The Johns Hopkins University Press.

Fondren, K. M. 2016. *Walking on the Wild Side.* New Jersey: Rutgers University Press.

Garshelis, D. L. and Pelton, M. R. 1980. Activity of Black Bears in the Great Smoky Mountains National Park. *Journal of Mammology* 61(1): 8–19.

Garshelis, D. L. and Pelton, M. R. 1981. Movements of Black Bears in the Great Smoky Mountains National Park. *The Journal of Wildlife Management* 45(4): 912–925.

Gibson, J. J. 1979. *The Ecological Approach to Visual Perception.* Boston: Houghton Mifflin.

Gore, M. L., Knuth, B. A., Curtis, P. D. and Shanahan, J. E. 2006. Stakeholder Perceptions of Risk Associated with Human–Black Bear Conflicts in New York's Adirondack Campgrounds: Implications for Theory and Practice. *Wildlife Society Bulletin* 34(1): 36–43.

Gore, M. L., Muter, B. A., Lapinski, M. K., Neuberger, L. and van der Heide, B. 2011. Risk Frames on Shark Diving Websites: Implications for Global Shark Conservation. *Aquatic Conservation: Marine and Freshwater Ecosystems* 21(2): 165–172.

Haraway, D. 2008. *When Species Meet.* Minneapolis: University of Minnesota.

Hill, J., Curtin, S. and Gough, G. 2014. Understanding Tourist Encounters with Nature: A Thematic Framework. *Tourism Geographies* 16(1): 68–87.

Hurn, S. 2012. *Humans and Other Animals.* London: Pluto Press.

Hurn, S. 2015. Baboon Cosmopolitanism. In: K. Nagai, K. Jones, D. Landry, M. Mattfeld, C. Rooney and C. Sleigh (eds.) *Cosmopolitan Animals*, pp. 152–166. Basingstoke: Palgrave Macmillan.

Ingold, T. 1994. Preface to the Paperback Edition. In: T. Ingold (ed.) *What Is an Animal?,* pp. xix–xxiv. London: Routledge.

Ingold, T. 2011. *Being Alive.* London: Routledge.

Jackson, M. 2002. *The Politics of Storytelling: Violence, Transgression, and Intersubjectivity.* Copenhagen: Museum Tusculanem.

Keul, A. 2013. Embodied Encounters between Humans and Gators. *Social & Cultural Geography* 14(8): 930–953.

Krakauer, J. 1996. *Into the Wild.* London: Pan Macmillan.

Ladino, J. K. 2009. For the Love of Nature: Documenting Life, Death, and Animality in *Grizzly Man* and *March of the Penguins. ISLE: Interdisciplinary Studies in Literature and Environment* 16(1): 53–90.

Landes, D. A. 2014. Translator's Introduction. In: M. Merleau-Ponty, *Phenomenology of Perception*, pp. xxx–li. Oxon: Routledge.

Laviolette, P. 2007. Hazardous Sport? *Anthropology Today* 23(6): 1–2.

Lorimer, J. 2015. *Wildlife in the Anthropocene.* Minneapolis: University of Minnesota Press.

Lyng, S. 1990. Edgework: A Social Psychological Analysis of Voluntary Risk Taking. *The American Journal of Sociology* 95(4): 851–886.

Merleau-Ponty, M. 2014 [1945]. *Phenomenology of Perception.* Oxon: Routledge.

Milton, K. 2002. *Loving Nature.* London: Routledge.

Milton, K. 2005. Anthropomorphism or Egomorphism? The Perception of Non-Human Persons by Human Ones. In: J. Knight (ed.) *Animals in Person: Cultural Perspectives on Human–Animal Intimacies*, pp. 255–271. Oxford: Berg.

Mitchell, M. S. and Powell, R. A. 2003. Response of Black Bears to Forest Management in the Southern Appalachian Mountains. *The Journal of Wildlife Management* 67(4): 692–705.

Nash, R. 1982. *Wilderness and The American Mind.* New Haven: Yale University Press.

NPS (National Park Service). 2016. Black Bears. Available at: www.nps.gov/grsm/ learn/nature/black-bears.htm.

Oelschlaeger, M. 1991. *The Idea of Wilderness.* New Haven: Yale University Press.

Orams, M. B. 2000. Tourists Getting Close to whales, Is It What Whale-Watching Is All About? *Tourism Management* 21: 561–569.

Peace, A. 2002. The Cull of the Wild. *Anthropology Today* 18(5): 14–19.

Pelton, M. R. 1972. Use of Foot Trail Travellers in the Great Smoky Mountains National Park to Estimate Black Bear (Ursus americanus) Activity. *Bears: Their Biology and Management* 2(23): 36–42.

Pelton, M. R., Scott, C. D. and Burghardt, G. M. 1976. Attitudes and Opinions of Persons Experiencing Property Damage and/or Injury by Black Bears in the Great Smoky Mountains National Park. *Bears: Their Biology and Management* 3: 157–167.

Philo, C. 1998. Animals, Geography, and the City: Notes on Inclusions and Exclusions. In: J. Wolch and J. Emel (eds.) *Animal Geographies: Place, Politics and Identity in the Nature–Culture Borderlands*, pp. 51–71. London: Verso.

Powell, R. A. and Seaman, D. E. 1990. Production of Important Black Bear Foods in the Southern Appalachians. *Bears: Their Biology and Management* 8: 183–187.

Quammen, D. 2003. *Monster of God.* London: W. W. Norton.

Reynolds, P. C. and Braithwaite, D. 2001. Towards a Conceptual Framework for Wildlife Tourism. *Tourism Management* 22: 31–42.

Reynolds-Hogland, M. J. and Mitchell, M. S. 2007. Effects of Roads on Habitat Quality for Bears in the Southern Appalachians: A Long-Term Study. *Journal of Mammology* 88(4): 1050–1061.

Reynolds-Hogland, M. J., Mitchell, M. S., Powell, R. A. and Brown, D. C. 2007. Selection of Den Sites by Black Bears in the Southern Appalachians. *Journal of Mammology* 88(4): 1062–1073.

Rogers, L. L. 1987. Effects of Food Supply and Kinship on Social Behaviour, Movements and Population Growth of Black Bears in Northeastern Minnesota. *Wildlife Monographs* 97: 1–72.

Rohini, C. K., Aravindan, T., Vinayan, P. A., Ashokkumar, M. and Anoop Das, K. S. 2016. An Assessment of Human–Elephant Conflict and Associated Ecological and Demographic Factors in Nilambur, Western Ghats of Kerala, Southern India. *Journal of Threatened Taxa* 8(7): 8970–8976.

Russell, C. L. 1995. The Social Construction of Orangutans: An Ecotourist Experience. *Society and Animals* 3(2): 151–170.

Schänzel, H. A. and McIntosh, A. J. 2010. An Insight into the Personal and Emotive Context of Wildlife Viewing at the Penguin Place, Otago Peninsula, New Zealand. *Journal of Sustainable Tourism* 8(1): 36–52.

Schutten, J. K. 2008. Chewing on the Grizzly Man: Getting to the Meat of the Matter. *Environmental Communication* 2(2):193–211.

Senthilkumar, K., Mathialagan, P., Manivannan, C., Jayathangaraj, M. G. and Gomathinayagam, S. 2016. A Study on the Tolerance Level of Farmers toward Human–Wildlife Conflict in the Forest Buffer Zones of Tamil Nadu. *Veterinary World* 9(7): 747–752.

Singer, F. J. and Bratton, S. P. 1980. Black Bear/Human Conflicts in the Great Smoky Mountains National Park. *Bears: Their Biology and Management* 4: 137–139.

Stokes, A. W. 1970. An Ethologist's Views on Managing Grizzly Bears. *BioScience* 20: 1154–1157.

Tate, J. and Pelton, M. R. 1983. Human–Bear Interactions in Great Smoky Mountains National Park. *Bears: Their Biology and Management*, Vol.5, A Selection of Papers from the Fifth International Conference on Bear Research and Management, Madison, Wisconsin, USA.

Urry, J. 1990. *The Tourist Gaze.* London: SAGE.

Veeder, B. 2016. *A Detailed Account of the Bear Attack at Spence Field.* Available at: https://peachpeak.wordpress.com/2016/05/24/first-blog-post/

Von Uexkull, J. 2010. *A Foray into the Worlds of Animals and Humans.* Minneapolis: University of Minnesota Press.

Williamson, J. F. Jr and Whelan, J. B. 1983. Computer-Assisted Habitat Mapping for Black Bear Management in Shenandoah National Park. *Bears: Their Biology and Management* 5: 302–306.

WNCN. 2016. *Wrong Bear Was Killed after Attack on NC Hiker, Officials Say.* Available at: http://wncn.com/2016/05/23/wrong-bear-was-killed-after-attack-on-nc-hiker-officials-say/

Wolch, J. 1998. Zoopolis. In: J. Wolch and J. Emel (eds.) *Animal Geographies: Place, Politics and Identity and the Nature–Culture Borderlands*, pp. 119–138. London: Verso.

Yarkovich, J., Clark, J. D. and Murrow, J. L. 2011. Effects of Black Bear Relocation on Elk Calf Recruitment at Great Smoky Mountains National Park. *The Journal of Wildlife Management* 75(5): 1145–1154.

3   Cuteness on the trail

Most interestingly, we encountered a raccoon on the trail after lunch. Did you know they can climb up trees? News to me. It looked adorable, its little face peeking around the trunk to see if we were still there.

The previous chapter looked at some of the different social constructions of black bears (*Ursus americanus*) by hikers, including one way of thinking about bears described as 'bear as Pet'. People who contributed to this narrative described the bears they encountered as "cute", "adorable", or even "tame". One blogger referred to a bear as "little buddy" and others talked about wanting to touch or "pet" a bear. The 'bear as Pet' narrative comes out of a way of reacting to bears that can be described as the 'cute response', in which the subject perceives the Other as vulnerable (see Angier, 2006; Dale et al., 2017; Ngai, 2012; Peplin, 2017; Richard, 2001), infant-like (see Buckley, 2016; Dale et al., 2017; Golle et al., 2013; Lorenz, 1971; Sanders, 1992), in need of parental care (see Aragon et al., 2015; Dale et al., 2017; Lorenz, 1971; Ngai, 2012; Serpell, 2003), adorable (see Ngai, 2012; Peplin, 2017; Richard, 2001), and endearing (see Angier, 2006; Dale, 2017; Gn, 2017).

Cute responses by hikers were not limited to bears on the trail; throughout their narratives bloggers described all kinds of species in terms of being cute, including ponies (*Equus ferus caballus*), pigs (*Sus scrofa domesticus*), raccoons (*Procyon lotor*), rabbits (*Lepus americanus* and *Lepus europaeus*), moose (*Alces alces*), mice (*Mus musculus*), turkeys (*Meleagris*), grouse (*Bonasa umbellus* and *Falcipennis canadensis*), box turtles (*Terrapene carolina*), beavers (*Castor canadensis*), marmots (*Marmota*), pine martens (*Martes americana*), and even rattlesnakes (*Crotalus*) and rat snakes (*Pantherophis obsoletus*). Indeed, rather than just small, furry animals being described as adorable, it seems that virtually any wild species could be perceived and described as cute by at least some of the people who encountered them. Where does this seemingly indiscriminate response by hikers to the other animals on the trail come from? Does it matter that at least some people responded to lots of the animals they encountered as being cute? Is the cute response as trivial, as harmless, as it might seem?

Since Konrad Lorenz's (1971) work looking at how people respond to baby-like features on human and other animal infants, the perception of cuteness has been understood as an innate or instinctive response on the part of the observer, yet Harris (2000: 5) writes that "cuteness...is not something we find in our children but something we *do* to them", placing the emphasis not on certain supposedly inherent characteristics displayed by an entity, but on a particular type of relationship between an entity and an observer. Harris's statement points to the possibility that we as observers may harbour certain motivations for relating to the Other, whether it be a human infant, another animal, or even an inanimate object, as being cute. This chapter will explore possible reasons that might have led to hikers seeing all kinds of animals on the trail as cute, and will argue that a hiker's perception of an animal as cute is more complex and meaningful to their overall experience of the animal than might initially be assumed. It will also address how a hiker seeing an animal as cute was meaningful for the animal as well, as it often affected how the hiker behaved towards them.

References to other-than-human animals as 'cute' or 'adorable' are pervasive in the current moment, in part due to the proliferation of cute memes on social media and other places online (see Albarran-Torres, 2017; Hutchinson, 2014; Lobato and Meese, 2014; Nekaris, 2013; Page, 2017; Peplin, 2017). It's possible that the cultural ubiquity of cuteness has something to do with the fact that only after studying hiker blogs for several months did I one day realise that hiker-bloggers were frequently writing about animals on the trail as cute, or referring to them in ways that either demonstrated that they had found the animal cute, or wanted to depict them as cute in their blogs. I soon found that I was not the only researcher not to notice cute-talk among research subjects; Granot et al. (2014: 67) write about how during their study of female consumers' retail decision making, they initially overlooked "cute", even though the word was used repeatedly by all of their participants. Something about cuteness, perhaps its pop-culture associations combined with a seeming shallowness, makes it recede into the background when looking for substantial data to explore.

Yet going back over my notes I found that just under a third of all hiker narratives mentioning wild animals on the AT indicated that the author had responded to the animal, either explicitly or more subtly, as being cute. A raccoon was described as "adorable...its little face peeking around the trunk", someone talked about having "an amusing interaction with a curious field mouse", and another blogger described coming across a wild pig, saying that they "fed it almonds and adored it a bit because it was cute". A hiker described a mother grouse trying to distract him from her infants by "scurrying [around]... like she was on top of a matchbox car". Another came across a "marmot that was adorable!" and another described pine martens as "very curious, friendly and fluffy". One blogger described how he had picked up a box turtle to get a closer look at him, saying "hey little guy", while others described woodland animals as "little critters". Occasionally narrators' references to animals on the trail as cute included a note of irony. One blogger, in a post titled "5 ways to avoid being eaten by bears", recommended patting a bear on the head and giving him a hug: "you could even tell him how cute he is…he's so fluffy I'm gonna die!" She also recommended taking a selfie with the bear, assuring the reader that the bear would want his picture taken. Another blogger described his impression of a bear's rear end.

> I mean, it's not like I expected it to be a sculpted work of art. But…I was kinda, sorta hoping it would be soft and cuddly, and maybe even do a cute side-to-side wiggle when it walked. Instead, it looked like two overly large Virginia smoked hams fighting over the remote control.

After re-reading hundreds of blog posts it became clear that seeing or talking about animals as being cute was a significant element to many hikers' experiences of the autonomous animals on the trail. From looking at older narratives by people describing encounters with wild animals (see for example Jones, 1867), it appears that responding to these types of species in a 'wilderness' setting as cute is a relatively new phenomenon, and one worth examining, despite the fact that cuteness has been described as "undeniably trivial" and "inconsequential" (Ngai, 2012: 18), "cheap" (Angier, 2006: np), and "banal" (Merish, 1996: 200).

The cute aesthetic

Harris (2000: 18) describes cuteness as "an aesthetic under siege, the object of contempt, laughter and scepticism". Yet the idea of cuteness wasn't always so derided. The origin of the cute aesthetic is found in ethology, when in 1943 Konrad Lorenz described a set of physical and behavioural characteristics that he referred to as innate "releasing schema for human parental care responses" (1971: 155). These included a relatively large head, prominent forehead, large eyes, bulging cheeks, short and thick limbs, and clumsy movements, which when combined gave the human or other animal "a loveable or 'cuddly' appearance" (1971: 154). Lorenz stated that when a human or nonhuman animal possessed the infant-like features he listed, the human observer of those features experienced an innate response that made them want to nurture and protect the bearer of those features which we would now refer to as 'cute'.

Since Lorenz, many other commentators have defined cute features as those that are especially juvenile or neotenic, including a diminutive size (especially in relation to conspecifics), chubbiness, softness, a tiny mouth, a button nose (see Buckley, 2016; Granot et al., 2014; Richard, 2001), and baby-like movements; "cute tumbles, toddles, waddles, rolls" (Richard, 2001: np). The appeal of something chubby and soft prompts Ngai to describe cuteness as "malleable", "amorphous", and even "bloblike" (2012: 30). Sanders, however, argues that "there is nothing essential about the link between any particular set of features and 'cuteness'; no set of features is intrinsically 'cute'. Rather, cuteness is just *any* set of features that is typical of babies" (1992: 163).

Angier has critiqued our supposed lack of discrimination when it comes to cute features.

> The human cuteness detector is set at such a low bar…that it sweeps in and deems cute practically anything remotely resembling a human baby or a part thereof, and so ends up including the young of virtually every mammalian species, fuzzy-headed birds like Japanese cranes, woolly bear caterpillars, a bobbing balloon, a big round rock stacked on a smaller rock, a colon, a hyphen and a close parenthesis typed in quick succession.
>
> (Angier, 2006: np)

In relation to nonhuman animals, fuzziness and furriness have been identified as being particularly cuddly (Granot et al., 2014), curly tails and floppy ears are supposedly adorable, (Dale et al., 2017), and the ability of a juvenile animal to maintain a vertical posture enhances his or her cuteness by making him or her appear more human-like and therefore easier to anthropomorphise (Mullan and Marvin, 1987). Tameness, or the willingness to approach and engage socially with the human, has also been identified as an important factor in nonhuman animal cuteness. Indeed, a certain set of perceived personality traits are deemed to be cute across the species, including vulnerability, fragility, innocence, helplessness, dependence, and playfulness (see Bryce, 2006; Granot et al., 2014; Richard, 2001), all of which also have an association with infancy. In criticising cuteness, Ngai writes that entities seem most cute when they "seem sleepy, infirm or disabled" (2012: 54), and Harris (2000) touches upon a growing association between the cute and the grotesque (see also Brzozowska-Brywczynska, 2007).

> Something becomes cute not necessarily because of a quality it has but because of a quality it lacks, a certain neediness and inability to stand alone, as if it were an indigent starveling, lonely and rejected because of a hideousness we find more touching than unsightly.
>
> (Harris, 2000: 4)

Ultimately, the cute aesthetic is made up of a largely agreed-upon combination of juvenile physical attributes, a way of moving through the world, and a perceived (correctly or incorrectly) status – vulnerable, needy, powerless. Together, these features are said to elicit certain prescribed types of reaction from human observers.

Responding to cuteness

If we assume that cuteness is trivial we may also conclude that our response to cuteness is trivial, or at least simple. However, there have been many reactions to cuteness reported, or what has variously been referred to as cute response or cuteness response (Dale, 2017; Serpell, 2003), cute-emotion (Buckley,

2016) or cute-affect (Dale et al., 2017), all phrases attempting to encompass the multi-faceted set of emotions that supposedly follow the perception of cuteness in an Other by a human being.

> Cute-emotion is a distinct and recognizable emotion, that complies with all the standard definitional criteria, such as characteristic hormonal responses, vocalizations, and facial expressions, recognizable across cultures and linked across senses...Its existence as an emotion, albeit as yet without a name, is recognized in psychological research.
>
> (Buckley, 2016: np)

Perhaps the least surprising reaction to cuteness is the rush of pleasure that many people experience when looking at something they perceive to be cute. Science writer Natalie Angier (2006: np) cites studies which have suggested that cute images stimulate the same pleasure centres of the brain that are aroused by "sex, a good meal or psychoactive drugs like cocaine". The easy hit of pleasure that supposedly comes with the experience of looking at something cute might explain not only the popularity of cute imagery on Facebook feeds, but also the use of cuteness in product marketing campaigns, and often in products themselves (see Albarran-Torres, 2017; Baker, 2001; Bryce, 2006; Goggin, 2017; Granot et al., 2014; Harris, 2000; Lawton, 2017; Ngai, 2012). Using cuteness to sell things not only works because people feel pleasure looking at the product or the advertisement for it, but also because people supposedly want to nurture the cute product, and to do so they need to purchase it first (Granot et al. talk about consumers "adopting" cute products, 2014: 69).

Since Lorenz described the human response to juvenile features as an urge to nurture, this has been the predominant narrative behind how we react to cuteness for the last half-century, but it has been recently challenged by psychologists Gary D. Sherman and Jonathan Haidt (2011), who argue that instead of releasing a protecting instinct, the cute response acts as a mechanism that releases *sociality*, i.e. the desire to play with and otherwise interact with the cute entity, which they argue is demonstrated by the kinds of behaviours that people exhibit when they perceive cuteness in an Other – attempts to touch, hold, play with, talk to, and otherwise interact with the cute entity.

> Cuteness is as much an elicitor of play as it is of care. It is as likely to trigger a childlike state as a parental one. There is almost certainly a relationship between cuteness and care, but we propose that the link is indirect and that the direct effect of cuteness is more general. We propose that the 'cuteness response' is an affective mechanism for detecting and responding to the social value of human children. As such, its primary *proximate* function is to motivate sociality, triggering attempts to engage the child in social interaction.
>
> (Sherman and Haidt, 2011: 4)

Sherman and Haidt continue by arguing that due to a proven link between sociality motivation and "mentalizing", which they describe as a process of attributing thoughts, beliefs, feelings, and intentions to an entity (although in the abstract to their paper "hyper-mentalizing" is explained as "humanizing", leaving nonhuman animals apparently outside the realms of thinking, believing, feeling or intending anything), perceiving cuteness in another entity should activate the perception of thoughts, beliefs, and feelings in that entity, and as a result, "cute entities become objects of moral concern, and members of the moral circle" (2011: 4). They conclude that it is thus possible for the cute-perceiving subject to feel compassion for the cute object.

Ngai (2012: 4) describes cute entities as evoking tenderness and an "unusually intense and yet strangely ambivalent kind of empathy"; yet for her, as for many other recent commentators on cute phenomena, the perception of cuteness and the supposed associated feelings of compassion or even pity for the cute entity have a dark side. Genosko (2005: np) concedes that cuddly entities create "emotional warmth", but argues that they do not elicit "understanding and respect". Ngai (2012: 54) describes the purported vulnerability and neediness of the cute entity as eliciting a desire in the observer to belittle or further diminish them, writing that "the experience of cute depends entirely on the subject's affective response to an imbalance of power between herself and the object". In other words, the pleasure that we feel when judging an Other as cute is bound up in an enjoyment of what we feel to be our higher status, and the entity's vulnerability *to us*. For Ngai, "cuteness is an aestheticization of powerlessness ('what we love because it submits to us')" (2012: 64).

Harris goes even further than Ngai, suggesting that cuteness aestheticises helplessness, and therefore "almost always involves an act of sadism on the part of its creator, who makes an unconscious attempt to maim, hobble and embarrass the thing he seeks to idolize" (2000: 5). Interestingly, Harris describes the subject perceiving cuteness in another entity as its "creator", reflecting the increasingly popular notion that cuteness is not actually inherent in an Other, but is something that we 'put onto' the Other (ostensibly for our own reasons) or "do to" the Other, as he puts it (2000: 5), in contradistinction to Lorenz's argument for innate releasing mechanisms (inherent cute features) in cute entities. In short, for Ngai and also Harris, the perception of and reaction to an entity as cute is an act of domination carried out upon an unwitting entity. This viewpoint comes close to Tuan's description of human affection for our companion animals – mainly dogs and cats – as an inescapably dominant attitude, indeed, as possible only *because of* our domination of them (1984), and may explain why we might generally expect the 'cute response' to be elicited by a companion animal rather than a wild (autonomous) animal.

Ngai also describes aggression as central to our experience of entities as being cute (2012: 23). Psychologists Aragon, Clark, Dyer, and Bargh (2015: 260) found that when they exposed their subjects to cute stimuli, the subjects experienced an impulse to nurture and care for the cute entity, and yet at the same time they also experienced an aggressive impulse, demonstrated by actions such as 'playful' growling, squeezing, biting, and pinching. The psychologists attributed these seemingly incongruous reactions to manifestations of "dimorphous expressions of emotion" (2015: 259), which they say occur when a person experiences one very strong emotion (for example, the urge to nurture a cute entity) and another emotion (for example, the aggressive response) emerges in order to provide a balance, and prevent the subject from being completely overwhelmed by the initial emotion. Another option proposed by Dyer is that the phenomenon known as "cute aggression" (Ferro, 2013: np) comes out of "high positive-affect, an approach orientation and almost a sense of lost control" (Ferro 2013: np). In effect, rather than the impulse to nurture being balanced by an aggressive impulse, the impulse to nurture is experienced so acutely that it tips over into frustrated aggression. Dale (2017: 41) argues that these feelings of aggression, rather than being directed at the cute entity, are actually directed back upon the subject (are masochistic, not sadistic), pointing to suddenly infantile behaviour of the subject "in the thrall of a cute object", and physiological responses such as tightened vocal cords, clenched fists and gritted teeth, which he describes as ways of discharging affective energy away from the cute entity.

The 'playful' squeezing, biting, and pinching that Aragon et al.'s subjects engaged in also demonstrates another reported effect of cuteness: the urge to touch the cute entity, or as Genosko puts it, "the need for contact with animals in the form of petting, patting or collecting [which] works itself out in a kindly aggressivity, a giddy proprietary right to possess, nurture, and display…" (2005: np). According to some, the need for contact can be so strong that it becomes a need to *consume* the cute object. The desire to consume cuteness can perhaps be more easily demonstrated when considering cute products, for example in their socio-marketing analysis of the concept of cuteness, Granot et al. write that "cute seems to be accessible *only* through consumption" (2014: 79, my emphasis), and de Vries talks about "a particular relationship between the edible and the adorable" (2017: 253), concluding that "devouring cuteness implies that we get to carry the cute around with us, make it *a part* of us" (2017: 271). In *Dominance and Affection: The Making of Pets* (1984) Tuan describes eating as an activity that takes pleasure in the object consumed, and is therefore an expression of love (1984: 9). He writes that "what we love we wish to incorporate, literally and figuratively" (1984: 9). Poltrack (2014: np) cites biological anthropologist Gwen Dewar, who attributes the urge to "nibble cute creatures" to a study that found that when women sniffed newborn infants the odours activated the same regions of the brain that experience pleasure from food. Poltrack also suggests that wanting to nibble a cute entity might be related to how other primates practice "social biting" on each other's infants (2014: np).

Finally, the "rapidity and promiscuity" (Angier, 2006: np) of the pleasurable, nurturing cute response means that people frequently find that initial impulse is followed by the sense that they are somehow being exploited, deceived, or "taken for a sucker", as philosopher Denis Dutton describes it (Angier, 2006: np). This again suggests that people seem to perceive their reaction to cuteness to be at least partly innate, even to the extent that they are frustrated by their own perceived involuntary response to cute stimuli. Dale describes how intentionality and agency can be drained from the subject by thinking about their response to cuteness in this manner.

> Quantitative analyses, which concentrate on affect, have isolated a specific set of stimuli that tends to cause a cuteness response in most people. However, this approach risks reducing cuteness to an autonomic, presubjective reaction that stands outside cognition, and thus apart from the production of meaning, the exercise of intention, and subjective agency in general.
>
> (Dale, 2017: 35)

A phenomenological experience of cuteness

Ngai comes close to a phenomenological way of thinking about cuteness, which reintroduces agency to the subject by thinking about the cute response as a way of looking.

> Softness, harmlessness, roundness, and so forth do not automatically give rise to the appearance of cuteness when combined. Richard Neer suggests that styles might therefore be understood as what Wittgenstein calls "aspects", "ways" of perceiving an object (seeing it *as* cute) as opposed to the set of objective qualities perceived.
>
> (Ngai, 2012: 29)

A reflexive approach to the cute response might conclude that it is not wholly automatic on the part of the subject, nor wholly learned (culturally acquired), nor chosen in some kind of vacuum in which the subject is able to decide upon their response free from all outside influences, but is a combination of biological makeup and social experiences, contingent on all kinds of influences from a subject's life interacting with each other. As Merleau-Ponty writes, "a thing is not actually *given* in perception, it is inwardly taken up by us, reconstituted and lived by us insofar as it is linked to a world whose fundamental structures we carry with ourselves" (2014 [1945]: 341). In *Biosocial Becomings* (2013) Ingold argues that "we can no longer think of the organism, human or otherwise, as a discrete, bounded entity, set over against an environment. It is rather a locus of growth within a field of relations traced out in flows of materials" (2013: 10). From this perspective, cuteness is not something that a cute entity does to the observing subject, nor something that the observing subject puts onto an entity, but something that happens between entities and within an environment.

Cute responses out on the trail

Many of the claimed effects of cuteness are evident in hiker narratives about their experiences with animals on the AT. Most commonly written about in terms of their adorableness were the wild ponies of the Grayson Highlands, on the Virginia portion of the trail.

> They were running and playing, manes flowing in the wind. Tiny foals were quite literally prancing back and forth between the legs of the adults. I fought the urge to clap my hands and squeal like a fourteen year old Japanese school girl. I understand now why there are so many signs reminding you that these are wild animals and should not be fed or petted, because as soon as I saw them it was all I could do to keep myself from chasing after them to pet and play... .
>
> This foal couldn't have been more than 4 feet high, walked right up and started licking my arm. I giggled. I smiled. Chills went up and down my spine. I was bad and went ahead and petted him anyways because at this point I simply could not restrain the urge any further. He nibbled on my arm at one point but it was totally worth it.
>
> Look, I'm a 34 year old man. I have a beard. I've changed the oil on my own car. I've gone hunting and fishing. I fit several stereotypes of manly, yet I could not get over just how gosh darn cute these things were.

This blogger's description of his encounter with highland ponies mentions many of the features commonly associated with cuteness: the particular ponies he focused on were the juveniles, and diminutive enough to weave between the legs of their parents, they were playful, and one approached the hiker and appeared to want to socialise with him, and to be touched by him. Many bloggers wrote about it being common knowledge that highland ponies would approach hikers for food (hikers often admitted to feeding them), and that ponies frequently licked or nibbled hikers because they enjoyed the saltiness of their sweat; as someone wrote, "I knew from…stories that the ponies liked licking salt off sweaty hikers, so I wasn't totally surprised". Accepting the common argument that the ponies' only interests were in food or salty sweat could have made hikers wary of the ponies' appeals, or even caused them to feel the sense of exploitation that Dutton describes (see Angier, 2006), but in general hiker-bloggers talked about having a similar reaction to this narrator's, feeling charmed because the ponies were cute to look at, and appeared to want to interact with them. Perhaps the fact that the ponies did want something from them even added to the cute appeal for many people; their perceived neediness put the hiker in the position of dominance considered integral to the experience of the adorable.

The narrator seems to revel in the strong cute response that he experienced in the presence of the ponies. He clearly took pleasure from his reaction to them, describing giggling and smiling despite himself. He was aware that he was not supposed to touch the pony that approached him, but apparently felt so overwhelmed by his cuteness that he simply couldn't help himself. To demonstrate just how strong his reaction to the ponies was, the narrator even describes getting chills from the experience.

By comparing his reaction to that of a "Japanese schoolgirl" and later by describing all of his "manly" attributes, such as having a beard, having changed the oil on his car, having gone hunting and fishing, the blogger makes clear his awareness of one of the implicitly understood features of the cute response; that it is a feminine way of perceiving certain entities (see Peplin, 2017; and on the pervasive popularity of cuteness (*kawaii*) in Japanese culture, see Allison, 2004; Kinsella, 1995; Miller, 2010). The association of the cute response with wanting to nurture and protect the cute entity makes it perhaps inevitable that the cute response be feminised. Some studies do appear to show that women are more sensitive to infant features than men (see Archer and Monton, 2010; Little, 2012), and there was some indication that certain male hikers were either less susceptible to cuteness, or wanted to show themselves as less affected by it, or experienced a different aspect of the response, as in the following extract from a blog posted by a female hiker.

> As we were walking down a narrow path, I spotted a tiny bunny nestled in the grass right next to the path. Roamer, a man who had recently started hiking with us, and sees himself as a bit of a forager, wanted to kill my little bunny for his next meal by hitting it on the head with a rock! Fortunately he refrained from doing so, but it unnerved me nonetheless!

However, these references to a possibly gendered reaction were minimal, with both males and females writing narratives about their experiences of the cute along the trail.

Why did hikers respond to wild animals as cute?

The identity work that is part of the process of telling stories about our experiences can influence how we choose to portray others featured in our narratives. A hiker's blog post can be constructed in a manner that helps them to make sense of the encounter, or to show themselves to their readers in a certain light. It is also a way for the hiker to demonstrate to readers that he feels at home enough in this environment to be able to intuitively understand the other animals that share the environment with him. Yet it seems that for many hikers, reacting to certain animals that they encountered on the AT as cute was a defining factor in their perception of those animals, as well as influencing how they behaved towards them. It should therefore be helpful to think about what may have prompted people to respond to animals on the trail as cute or loveable.

It would be easy to focus on the false dichotomy of trying to decipher where the cute response comes from: is it biological? Is it socially learned?

Yet, as Ingold says, "there is no division between them. The domains of the social and the biological are one and the same" (2013: 9). The perception of cuteness is a particularly good example of what Ingold terms the "biosocial". At least to a degree, it appears that Konrad Lorenz was right, and that human beings do respond to certain characteristics that signal that a human or other animal is an infant. Recent studies have appeared to support Lorenz's argument, including one by Archer and Monton (2010) in which their subjects showed a preference for faces with infant features over those without; one by Little, who concluded that his data "demonstrate that baby-like traits have a powerful hold over human perceptions" (2012: 775), and Golle et al.'s study of perceptual adaptation to "babyfacedness" transferring across species, which they stated suggested a "common mechanism coding cuteness in human and nonhuman faces" (2013: 1). Aragon et al. (2015) successfully used cute stimuli to generate intense positive emotion in study participants, as a precursor for demonstrating their theory that cuteness was an elicitor of dimorphous expressions of emotion. Thus, it is possible to argue that hikers, to some degree at least, may have been responding to certain characteristics that some of the animals on the trail possessed. However, the specific 'innate releasing schema' – the cute and cuddly features that Lorenz argued release nurturing and sociality impulses in humans – were not possessed by all the animal species that hikers responded to as being cute, and so clearly there were also other influences at work.

For many, if not most of the hiker-bloggers on the AT, thru-hiking the trail is the first time that they have had contact with numerous different animal species. Thus, their perception of animal life on the trail as cute could be at least partly attributed to the influence of cultural representations of the kinds of animals that make up the category 'wildlife', which may include Disney films (see Wilson, 1992), cartoons, the types of nature documentaries that portray animals as being cute or lovable (see Candea, 2010), pictures on breakfast cereal packaging, sports mascots, totem-like tattoos, toys (see Klaffke, 2012), and, of course, cute advertising (see Baker, 2001), which seems particularly unavoidable these days. Cuteness is also ubiquitous online, where cute memes of all types of animal species proliferate on social media, from puppies and bears, to geckos, slugs, and snakes (see Gall Myrick, 2015; Hutchinson, 2014; Laforteza, 2014; Lobato and Meese, 2014; Meese, 2014; Miltner, 2014; O'Meara, 2014; Page, 2017; Peplin, 2017; Potts, 2014; Steinbock, 2017; Wittkower, 2012). Nekaris et al. (2013) looked at user comments in response to a YouTube video of a slow loris being tickled, and found that one of the top three most common comments by viewers of the video was how cute they thought the slow loris was, and that one in ten commenters wrote that they wanted one as a pet, perhaps demonstrating the proprietary nature of the cute response. It's also worth noting that cute-culture is so pervasive online that the online setting of the Trek (www.thetrek.co) may possibly have had an influence on the extent to which hikers talked about animals on the trail as being cute. Indeed, Wittkower has argued that cuteness is a "dominant aesthetic of digital culture" (cited in Albarran-Torres, 2017: 243). According to Baker, "popular culture...sees only itself in the eyes of its animal" (2001: xxi).

Also culturally pervasive at the moment is the idea of wilderness as benefi-cent and welcoming. This was explored in Chapter 1, in looking at how hiker approaches to their wilderness trek as a pilgrimage influenced how they saw the animals on the trail, who, as the inhabitants of the wilderness, were often viewed as equally beneficent and welcoming. It is possible to react to this idea of wildlife with awe and reverence, emotions that some people reported feeling in the presence of nonhuman animals on the trail. However, the 'ben-eficent wildlife' attitude also makes it easier to respond to wild animals as cute and cuddly, given that the inhabitants of a welcoming wilderness are less likely to be viewed as dangerous, unpredictable or transgressive. Dale et al. (2017: 5) write that "the desire to enter, if only for a moment, a state of being that renders the world unthreatening and playful comprises a compel-ling link between the aesthetics of cute and the cute affect". As mentioned in Chapter 1, many hikers described their reasons for hiking the trail as an escape from what they perceived to be the ills of modern society, and Dale et al. describe animals as "serving as facilitators for safe forms of socializa-tion and enabling us to attribute to them a set of positive values" that might be perceived to be missing from the day-to-day social landscape (2017: 21). They go on to describe animals as often being "enlisted for... empathic sus-tenance and consolation" (2017: 23) by people struggling with the pressure of inhabiting an ever-changing "risk" society. Yet believing that the animal is there for the hiker's benefit and enjoyment, to enhance their wilderness experi-ence, to relieve feelings of pressure or alienation, or to appear as a highlight in the hiker-blogger's photographs or narratives is a diminishment of the animal. It is perhaps particularly problematic when thinking about wild animals.

> A modern man may claim intimacy with nature – with wilderness itself. But this sense of ease in wilderness is possible only because wild animals and forests are no longer threatening. Wilderness, although not yet a pet...nonetheless is widely perceived by modern society to be a fragile existence that needs its protection.
>
> (Tuan, 1984: 163)

This leads on to another common feature of our cultural lives: the assumption of human domination over all animal life – the notion that defines the "Anthropocene" (see Crutzen, 2002). Taking for granted that we have con-trol over the animal Other puts us in a position of power over them (real or imagined), a position integral to the perception of other entities as cute. Washbaugh and Washbaugh (2000) cite the apparently quite popular practice of fishermen kissing trout before releasing them as a simultaneous display of love and domination of nature. When the fish is netted, the angler bends down to the water, "planting one's lips on the trout as it lies there near the water, half submerged in one's catch-and-release net, held with a well-moistened hand gently cradling her body" (2000: 124). Nobody is suggesting that the angler finds the trout to be cute, but it seems to be that the position of total control over the fish in the moment in which she is caught enables the angler to be filled with affection for her.

Alternatively, an underlying feeling of *loss* of power could also be responsible for a strong cute response in other people. Sherman and Haidt (2011), who argued that the perception of cuteness leads to 'mentalisation' of the cute entity, cite the research of Epley et al. (2007, 2008), who focus on what they describe as a particular form of mentalising: anthropomorphisin g. Milton (2005: 255) describes anthropomorphising as "the attribution of human characteristics to non-human things". Epley et al. (2007: 2) "propose that people are most likely to anthropomorphize when they are particularly motivated to reduce unpredictability or otherwise gain a sense of control". This would suggest that rather than responding to animals as cute because they feel in control of the encounter, hikers might respond to animals as cute in an attempt to regain a feeling of control that they are lacking. Most of the hikers blogging had never ventured into a wilderness setting before, and some reported feeling overwhelmed or alienated by the new, unfamiliar environment ("on my first night out here I was shivering in my tent, terrified"). A response to the animals on the trail as being cute (and the so-called mentalising that they might engage in as part of the response) could potentially be viewed almost as a means of self-reassuring, a way of seeing an environment that could be felt as alien and hostile as friendly and non-threatening instead.

In summary, then, factors influencing hikers' biosocial experiences of animals on the trail as cute could include so-called innate responses to infant-like features, the influence of cute-culture on their perception of wild animals, the notion of the wilderness as beneficent and nonthreatening, assumptions of dominance, or the opposite, and feelings of powerlessness that lead hikers to seek reassurance in the lovable-ness of other animals on the trail. In addition, hikers may have been motivated to portray animals as cute in their blogs in order to say something about themselves (dominant and in control, or nurturing and kind), or because this is a fairly standard way for animals to be depicted online.

The effect of seeing animals as cute

Does seeing autonomous animals as cute *matter*? For Sherman and Haidt (2011), the idea that people 'mentalise' the entities that they find cute increases their possibility for social connection with those entities. From this perspective, hikers reacting to other animals as cute would increase the likelihood of them being able to engage socially with those animals, which in turn could presumably help them to forge a better appreciation of them as individuals. Yet Harris sees the *opposite* happening, arguing that "the cute vision of the natural world is a world without nature, one that annihilates 'otherness', ruthlessly suppresses the nonhuman" (2000: 12). This view of the cute response suggests that by responding to another animal as cute, we make him into what we want or need him to be, rather than seeing what he *is*. Merish goes even further than Harris, describing cuteness as a seeming disavowal of otherness, which assimilates the Other into middle class "familial and emotional structures", and which transforms "transgressive subjects into beloved objects" (1996: 194). Peplin seems to disagree.

> It is the ongoing relationship of care that truly marks the cute as an object different from the pitied, the weak, or the small; a cute object must be swooned over, cared for, worried about; it becomes a subject of imaginative, if not actual, care...To be cute is to be weak, small, and powerless, yes, but in a way that demands a relationship between the subject and the object.
>
> (Peplin, 2017: 115)

In hiker narratives about perceiving an animal on the trail as cute, their response was frequently accompanied by behaviour that had an effect on the life of the animal – which created the relationship between subject and object that Peplin talks about. A common response was to stop and watch the animal, although some bloggers talked about making sure not to come too close, or staying as still as they could in order not to scare him or her away. For many animals, the awareness of being watched will have been an uncomfortable experience, although hikers either did not know this, or did not acknowledge it in their blogs. Many bloggers talked about stopping to take photos of the cute animal, with some describing approaching as close as they could in order to do so, particularly those who wanted to take 'selfies' with the animals. For example, one blogger talked about encountering a young deer, saying: "once she realized I wasn't a threat, she just chilled and allowed me to get a snapchat vid of her and a couple pics".

Peace (2005: 205) describes Western culture as having an "ocularcentric quality" and even a "visual obsession". Garland-Thomson (2009: 9) describes the gaze as "an oppressive act of disciplinary looking that subordinates its victim". Her focus is on the meaning behind one human staring at another, but her argument is equally applicable to how a human stares at a nonhuman.

> Cultural othering in all its forms – the male gaze being just one instance – depends upon looking as an act of domination. The ethnographic or the colonizing look operates similarly to the gendered look, subordinating its object by enacting a power dynamic. When persons in a position that grants them authority to stare take up that power, staring functions as a form of domination, marking the staree as the exotic, outlaw, alien or other. The colonizing look marks its bearer as legitimate and its object as outsider.
>
> (Garland-Thomson, 2009: 42)

What Franklin refers to as the "natural zoological gaze" (1999: 81) means that, for Berger, "animals are always the observed…They are the objects of our ever-extending knowledge. What we know about them is an index of our power…" (2009: 16). It is thus not possible to argue that looking (without interfering) at cute animals is not an intrusion into their lives, or is not an act of domination. Peace's research into whale watching led him to conclude that it "involves the extensive consumption of the symbolism or the totemism of the whale" (2005: 194) and, likewise, watching cute animals in the woods can be regarded as the consumption of them as metonyms of 'wilderness', particularly when it comes to taking photos and "snapchat vids", and claiming that the animal was "chilled" about the situation.

Many hiker narratives about cute animals described their attempts at interacting with the animal, either through making 'animal-like' noises in an effort to communicate, or more regularly, through trying to touch the animal. Some attempts at touching seem to have been quite cautious, like holding a hand out for the animal to decide if she wanted contact, yet most described "petting" the animal. This seemed to occur most often with highland ponies, who it was commonly said, *wanted* to be petted. One hiker described her disappointment that the ponies she met didn't seem to be interested in being stroked.

[T]he 3 we saw were the most apathetic ponies in the world. Other hikers talk about how they pet, fed and took selfies with the ponies. Meanwhile, I was barely able to pet a pony via my trekking pole. It just kept edging away from me.

Tuan (1984: 171) describes petting as the enactment of a relationship of dominance and subordination, which is reinforced by the subject feeling free to reach down and stroke the object. Although he describes this action as a gesture of affection, he notes that these types of gestures "are bestowed by the superior on the inferior and can never be used between equals" (1984: 171). He writes similarly about feeding an animal, which was also popular, with hikers describing feeding or trying to feed ponies, a pig, shelter mice, and chipmunks, amongst others. Other attempts at 'nurturing' included a hiker who crafted a nest for a mouse and her newborn babies, "placing the infants by the shelter" (presumably so that they could find it). Stroking and feeding animals, or building a nest for a mouse mother and her infants, are indicative of the occasionally strong feelings of care that accompanied hiker responses to animals as being cute; they tried to find ways of caring for the animals, even in inappropriate circumstances (for example, feeding carrots to ponies while fully aware of the signs requesting not to do this, or trying to stroke a pony even when the pony resisted). In accordance with Peplin's argument, these animals became subjects of "imaginative, if not actual, care" (2017: 115).

Yet at least some of the encounters described will have been frightening for the animals involved, including a hiker who described chasing a grouse up the trail accompanied by her hiking companion ("Get her!"), and another who came across a box turtle.

> He caught my eye and I stopped dead in my tracks just a foot in front of him. I crouched down to get a closer look and he immediately retracted into his shell, as if he were terrified of my presence. I picked him up, probably against his will, to inspect him further. It was like I was a child for a moment, in awe of something so relatively trivial...

Conclusion

This chapter has examined how hikers on the AT responded to, and wrote about, animals on the trail as cute. Considering the generally accepted aesthetics of cuteness, as well as the range of emotions that are associated with the cute response, has helped in thinking about why hikers on the trail perceived animals as adorable, and how that perception informed their experiences with animals. Far from being trivial or shallow, the perception of cuteness in other animals is significant, particularly as this cute response frequently had a direct impact on the life of the animal who was perceived that way.

All ways of representing animals matter, even, or perhaps especially, the ones that might seem banal or meaningless. Baker, talking about representations of animals in general, writes that it is important to "inquire into the consequences of their apparent inconsequentiality" (2001: 3). In continuing to think in dualisms, in the ongoing uncertainty about whether the cute response is innate or learned, whether cuteness is inherent in an object, or inflicted on the object by an observing subject, the cute response has become something like a myth, which, says Barthes, "transforms history into nature" (2000 [1957]: 129), essentially eradicating the influence of history and culture on an idea, and presenting it as 'only natural'. Yet the best example of this is not the cute response itself, but the myth of cuteness as inconsequential, the idea that the perception of an animal as cute is shallow and meaningless, and the attitude that animals who appear cute are themselves trivial.

> Myth does not deny things, on the contrary, its function is to talk about them; simply, it purifies them, it makes them innocent, it gives them a natural and eternal justification, it gives them a clarity which is not that of an explanation but that of a statement of fact...it organizes a world which is without contradictions because it is without depth, a world wide open and wallowing in the evident, it establishes a blissful clarity...
>
> (Barthes, 2000 [1957]: 143)

Cuteness, portrayed as simple and self-evident, need not, therefore, be looked at for any depth of meaning. Baker (2001) cites Barthes' description of myth, noting that, for Barthes, myth functions to make things seem unworthy of analysis. Baker points to the "unwritten priorities" of our culture which effectively drain certain things of their significance. He writes that "the dominant cultural view that the subject of animals is essentially trivial...is a clear case in point" (2001: 8). A particularly effective way to portray animals as insignificant is to portray them as cute, and concurrently to imply that cuteness is simple, childish, feminine, unimportant. Interestingly, Baker himself dismisses cuteness when he talks about certain types of art featuring animals, saying that he "took rather a dim view of such art, associating it with cuteness and kitsch" (2001: xxvi). The myth of cuteness as inconsequential was at work even here; for Baker these types of art seemed, at least initially, unworthy of analysis.

Cuteness as an aesthetic, as a way of looking, as a response that can elicit a multitude of different emotions and as a tool in storytelling, is not inconsequential at all. Responding to animals as cute was a large part of how many hikers perceived and experienced the animals they met on the trail. Their responses frequently also had an impact on the lives of those animals, as human-perceived cuteness in animals has been shown to do, both on an individual and a species level. Stephen Budiansky (1992) has described how in the early stages of domestication unconscious human preferences for juvenile features may well have reinforced a selection for those neotenic traits, influencing the lives and future lives of those species who began to live closely with humans. In an age of extinctions, biologist David L. Stokes (2006) has argued that human aesthetic preferences among species could actually determine which species survive and which do not. Finding a species attractive or cute can alert the public's attention to their cause, for example in the case of the giant panda (see Baker, 2001: 178–180). Yet it can also be detrimental. Nekaris et al. (2013) include capture for the photo prop and pet trades as one of the major threats to the survival of slow lorises, who are currently threatened with extinction. Animals on the AT who were perceived as cute frequently had their personal space intruded upon, were watched, photographed, stroked, picked up, fed, followed, and chased. Some of these situations will have been frightening, and some, such as being able to approach people for food, could be seen as an 'advantage' of cuteness.

In all sorts of ways, embodied encounters with animals on the AT gave hikers a deeper insight into the lives of the beings that they met, while at the same time hikers never fully moved away from the idea of the animal *meaning something* in the context of their own journey on the trail. While the cute response certainly has some negative connotations, when hikers responded to animals on the trail as being cute they often became very engaged with the animal, and with what the animal was doing and how she was responding to them, and this interested engagement with the animal can be seen as opening up possibilities for learning about her, her personhood, and her life-world. Feelings of care were often elicited by reactions to animals as being cute. Van Dooren (2014: 294) describes the act of caring as an "embodied phenomenon", and argues that "caring is always a practice of worlding" (2014: 294), pointing to the possibility that hiker demonstrations of caring towards animals on the trail could be viewed as such acts of "worlding", embedding themselves in place with the inhabitants of the trail. At the same time, the animal's perceived cuteness could be interpreted as carrying meaning, for example, that the wilderness is indeed a friendly place, that the hiker is welcome there, that the hiker has nothing to fear from the unfamiliar environment, and that the hiker is having the life-affirming experience that he walked onto the trail looking for.

References

Albarran-Torres, C. 2017. Kittens, Farms and Wild Pandas: The Impact of Cuteness in Adult Gamble-Play Media. In: J. P. Dale, J. Goggin, J. Leyda, A. P. McIntyre and D. Negra (eds.) *The Aesthetics and Affects of Cuteness*, pp. 235–252. London: Routledge.

Allison, A. 2004. Cuteness as Japan's Millennial Product. In: J. J. Tobin (ed.) *Pikachu's Global Adventure*, pp. 34–49. Durham: Duke University Press.

Angier, N. 2006. The Cute Factor. *New York Times*. Available at: www.nytimes.com/2006/01/03/science/the-cute-factor.html

Aragon, O. R., Clark, M. S., Dyer, R. L. and Bargh, J. A. 2015. Dimorphous Expressions of Positive Emotion: Displays of Both Care and Aggression in Response to Cute Stimuli. *Psychological Science* 26(3): 259–273.

Archer, J. and Monton, S. 2010. Preferences for Infant Facial Features in Pet Dogs and Cats. *Ethology* 117: 217–226.

Baker, S. 2001. *Picturing the Beast*. Champaign: University of Illinois Press.

Barthes, R. 2000 [1957]. *Mythologies*. London: Random House.

Berger, J. 2009 [1980]. *About Looking*. London: Bloomsbury.

Bryce, M. 2006. Cuteness Needed: The New Language/Communication Device in a Global Society. *International Journal of the Humanities* 2(3): 2265–2275.

Brzozowska-Brywczynska, M. 2007. Monstrous/Cute. Notes on the Ambivalent Nature of Cuteness. In: N. Scott (ed.) *Monsters and the Monstrous. Myths and Metaphors of Enduring Evil*, pp. 213–228. Amsterdam: Rodopi.

Buckley, R. C. 2016. Aww: The Emotion of Perceiving Cuteness. *Frontiers in Psychology* 7: Article 1740. Available at: www.frontiersin.org/articles/10.3389/fpsyg.2016.01740/full

Budiansky, S. 1992. *The Covenant of the Wild: Why Animals Chose Domestication*. New York: W. Morrow.

Candea, M. 2010. "I Fell in Love with Carlos the Meerkat": Engagement and Detachment in Human–Animal Relations. *American Ethnologist* 37(2): 241–258.

Crutzen, P. J. 2002. Geology of Mankind. *Nature* 415: 23.

Dale, J. P. 2017. The Appeal of the Cute Object: Desire, Domestication and Agency. In: J. P. Dale, J. Goggin, J. Leyda, A. P. McIntyre and D. Negra (eds.) *The Aesthetics and Affects of Cuteness*, pp. 35–55. London: Routledge.

Dale, J. P., Goggin, J., Leyda, J., McIntyre, A. P. and Negra, D. (eds.). 2017. *The Aesthetics and Affects of Cuteness*. New York: Routledge.

van Dooren, T. 2014. Care. *Environmental Humanities* 5: 291–294.

Epley, N., Waytz, A. and Cacioppo, J. T. 2007. On Seeing Human: A Three-Factor Theory of Anthropomorphism. *Psychological Review* 114: 864–886.

Epley, N., Akalis, S., Waytz, A. and Cacioppo, J. T. 2008. Creating Social Connection Through Inferential Reproduction: Loneliness and Perceived Agency in Gadgets, Gods and Greyhounds. *Psychological Science* 19: 114–120.

Ferro, S. 2013. Why Do We Want to Squeeze Cute Things? *Popular Science.* Available at: www.popsci.com/science/article/2013-01/science-says-adorable-animals-turn-us-aggressive

Franklin, A. 1999. *Animals and Modern Cultures.* London: SAGE Publications Ltd.

Gall Myrick, J. G. 2015. Emotion Regulation, Procrastination, and Watching Cat Videos Online: Who Watches Internet Cats, Why, and to What Effect? *Computers in Human Behavior* 52: 168–176.

Garland-Thomson, R. 2009. *Staring: How We Look.* New York: Oxford University Press.

Genosko, G. 2005. Natures and Cultures of Cuteness. *Invisible Culture: An Electronic Journal for Visual Culture* 9. Available at: www.rochester.edu/in_visible_culture/Issue_9/genosko.html

Goggin, J. 2017. Affective Marketing and the Kuteness of Kiddles. In: J. P. Dale, J. Goggin, J. Leyda, A. P. McIntyre and D. Negra (eds.) *The Aesthetics and Affects of Cuteness*, pp. 216–234. London: Routledge.

Golle, J., Lisibach, S., Mast, F. W. and Lobmaier, J. S. 2013. Sweet Puppies and Cute Babies: Perceptual Adaptation to Babyfacedness Transfers across Species. *Plos One* 8(3): 1–5.

Granot, E., Alejandro, T. B. and Russell, L. M. 2014. A Socio-Marketing Analysis of the Concept of Cute and Its Consumer Culture Implications. *Journal of Consumer Culture* 14(1): 66–87.

Harris, D. 2000. *Cute, Quaint, Hungry and Romantic.* New York: Basic Books.

Hutchinson, J. 2014. I Can Haz Likes: Cultural Intermediation to Facilitate "Petworking". *M/C Journal* 17(2). Available at: http://journal.media-culture.org.au/index.php/mcjournal/article/view/792

Ingold, T. 2013. Prospect. In: T. Ingold and G. Palsson (eds.) *Biosocial Becomings: Integrating Social and Biological Anthropology*, pp. 1–21. Cambridge: Cambridge University Press.

Jones, J. M. 1867. A Fortnight in the Backwoods of Shelburne and Weymouth. Available at: dalspace.library.dal.ca/handle/10222/10158

Kinsella, S. 1995. Cuties in Japan. In: B. Moeran and L. Scov (eds.) *Women, Media and Consumption in Japan*, pp. 220–254. London: Curzon Press.

Klaffke, P. 2012. *Hello Cutie!* Vancouver: Arsenal Pulp Press.

Laforteza, E. M. 2014. Cute-ifying Disability: Lil Bub, the Celebrity Cat. *M/C Journal* 17(2). Available at: http://journal.media-culture.org.au/index.php/mcjournal/article/view/784

Lawton, L. 2017. Taken by the Tamagotchi: How a Toy Changed the Perspective on Mobile Technology. *The iJournal* 2(2). Available at: http://theijournal.ca/index.php/ijournal/article/view/28127

Little, A. C. 2012. Manipulation of Infant-Like Traits Affects Perceived Cuteness of Infant, Adult and Cat Faces. *Ethology* 118: 775–782.

Lobato, R. and Meese, J. 2014. Kittens All the Way Down: Cute in Context. *M/C Journal* 17(2). Available at: http://journal.media-culture.org.au/index.php/mcjournal/article/view/807

Lorenz, K. 1971. *Studies in Animal and Human Behaviour, Volume 2*. London: Methuen & Co.

Meese, J. 2014. "It Belongs to the Internet": Animal Images, Attribution Norms and the Politics of Amateur Media Production. *M/C Journal* 17(2). Available at: http://journal.media-culture.org.au/index.php/mcjournal/article/view/782

Merleau-Ponty, M. 2014 [1945]. *Phenomenology of Perception*. Oxon: Routledge.

Merish, L. 1996. Cuteness and Commodity Aesthetics. In: R. G. Thomson (ed.) *Freakery: Cultural Spectacles of the Extraordinary Body*. New York: New York University Press.

Miller, L. 2010. Japan's Zoomorphic Urge. *ASIANetwork Exchange* XVII(2): 69–82.

Miltner, K. 2014. "There's No Place for Lulz on LOLCats": The Role of Genre, Gender, and Group Identity in the Interpretation and Enjoyment of an Internet Meme. *First Monday* 19(8). Available at: http://firstmonday.org/ojs/index.php/fm/article/view/5391/4103

Milton, K. 2005. Anthropomorphism or Egomorphism? The Perception of Non-Human Persons by Human Ones. In: J. Knight (ed.) *Animals in Person: Cultural Perspectives on Human–Animal Intimacies*, pp. 255–271. Oxford: Berg.

Mullan, B. and Marvin, G. 1987. *Zoo Culture*. London: Weidenfeld & Nicolson.

Nekaris, A., Campbell, N., Coggins, T. G., Rode, E. J. and Nijman, V. 2013. Tickled to Death: Analysing Public Perceptions of "Cute" Videos of Threatened Species (Slow Lorises – *Nycticebus* spp.) on Web 2.0 Sites. *Plos One* 8(7): 1–10.

Ngai, S. 2012. *Our Aesthetic Categories: Zany, Cute, Interesting*. Cambridge: Harvard University Press.

O'Meara, R. 2014. Do Cats Know They Rule YouTube? Surveillance and the Pleasures of Cat Videos. *M/C Journal* 17(2). Available at: http://journal.media-culture.org.au/index.php/mcjournal/article/view/794

Page, A. 2017. "This Baby Sloth Will Inspire You to Keep Going": Capital, Labour, and the Affective Power of Cute Animal Videos. In: J. P. Dale, J. Goggin, J. Leyda, A. P. McIntyre and D. Negra (eds.) *The Aesthetics and Affects of Cuteness*, pp. 75–94. London: Routledge.

Peace, A. 2005. Loving Leviathan: The Discourse of Whale-Watching in Australian Ecotourism. In: J. Knight (ed.) *Animals in Person*, pp. 191–210. New York: Berg.

Peplin, K. 2017. Live Cuteness 24/7: Performing Boredom on Animal Live Streams. In: J. P. Dale, J. Goggin, J. Leyda, A. P. McIntyre and D. Negra (eds.) *The Aesthetics and Affects of Cuteness*, pp. 112–129. London: Routledge.

Poltrack, E. 2014. Why Do We Want to Bite Cute Things, Like Adorable Newborn Babies? *Scientific American*. Available at: www.scientificamerican.com/article/why-do-we-want-to-bite-cute-things-like-adorable-newborn-babies/

Potts, J. 2014. The Alchian–Allen Theorem and the Economics of Internet Animals. *M/C Journal* 17(2). Available at: http://journal.media-culture.org.au/index.php/mcjournal/article/view/779

Richard, F. 2001. Fifteen Theses on the Cute. *Cabinet 4*. Available at: www.cabinetmagazine.org/issues/4/cute.php

Sanders, J. T. 1992. On "Cuteness". *British Journal of Aesthetics* 32(2): 162–165.

Serpell, J. A. 2003. Anthropomorphism and Anthropomorphic Selection – Beyond the "Cute Response". *Society & Animals* 11(1): 83–100.

Sherman, G. D. and Haidt, J. 2011. Cuteness and Disgust: The Humanizing and Dehumanizing Effects of Emotion. *Emotion Review* 3(3): 1–7.

Steinbock, E. 2017. Catties and T-Selfies. *Angelaki* 22(2): 159–178.

Stokes, D. L. 2006. Things We Like: Human Preferences among Similar Organisms and Implications for Conservation. *Human Ecology* 3(35): 361–369.

Tuan, Y. 1984. *Dominance and Affection: The Making of Pets*. New Haven: Yale University Press.

De Vries, N. 2017. Under the Yolk of Consumption: Re-Envisioning the Cute as Consumable. In: J. P. Dale, J. Goggin, J. Leyda, A. P. McIntyre and D. Negra (eds.) *The Aesthetics and Affects of Cuteness*, pp. 253–273. London: Routledge.

Washbaugh, W. and Washbaugh, C. 2000. *Deep Trout: Angling in Popular Culture*. London: Bloomsbury Academic.

Wilson, A. 1992. *The Culture of Nature*. Oxford: Blackwell Publishers.

Wittkower, D. E. 2012. On the Origins of Cute as a Dominant Category in Digital Culture. In: T. Luke and J. Hunsinger (eds.) *Putting Knowledge to Work and Letting Information Play*, pp. 212–221. Boston: Springer.

4   Ugly, scary and disgusting

Uncomfortably close encounters with mice, snakes, insects, and other 'critters' on the trail

This chapter looks at hiker narratives that described unpleasant encounters between themselves and other animals on the trail. Jerolmack (2008: 89) writes that "wild or feral animals move about with their own trajectories". When hiker trajectories collided with those of certain types of animals – most often mice, snakes and various insects, but also worms, leeches, skunks, and more – these collisions would often result in uncomfortable encounters, what Jerolmack describes as "an underlying existential human experience of social disorder" (2008: 89). Chapter 2 explored uncomfortable encounters with bears, so they won't be included here. Furthermore, unlike bears, what the animals featured in this chapter have in common is that they are generally not the types of species that hikers fantasised about meeting on their thru-hike. Many of them represent radical Otherness, forms of otherness that would seem to preclude the possibility for the kind of communicative encounters that hikers talked about having with bears, ponies, various small mammals, and even birds. The animals encountered in this chapter were largely viewed as 'pests' by hiker-bloggers, whose narrations about them contributed to the existing pestilence discourse (see Knight, 2000) around certain animals on the trail. Some of these encounters were simply uncomfortable or disconcerting, while in others hikers described themselves as "terrified", "repulsed", and "freaked out". Yet there is also an element of curiosity and fascination with which many of them are narrated, demonstrating an interesting 'push–pull' effect that animals described as disgusting or terrifying had on some people. While many hikers spoke of running or moving away from these species in the moment of their encounter, their subsequent blog posts often described in careful, intricate detail the way the animal looked (their sliminess or hairiness, the size of their feet, the thickness of their body) and moved (crawling, sprinting, pitter-pattering or slithering), demonstrating that although people felt motivated to end uncomfortable encounters quickly, they often relished thinking about and describing the animal afterwards.

What was it like to live in close proximity to the types of animals that people usually avoid? What about them meant that they were perceived as distasteful, disgusting or frightening? How did hikers respond to these animals in the moment, and what were the consequences of the hiker's response for the animal? Finally, how did hikers think about these animals and make sense of their encounters with them in their blog posts?

Disgust and animals

> I wanted to experience sleeping in a shelter at least once, so I set up my stuff inside the three-sided structure…Suddenly I discovered in myself a serious fear of mice, of their needle-ish claws, their pittering, scampering feet, their long, bald tails, and the filthy microorganisms that fester in their dun coats. Repulsed, I fled to set up my tent behind the shelter…

Hikers experienced a range of emotions in their uncomfortable encounters with animals, including fear, confusion, irritation, curiosity, and a kind of fascinated horror. Of the negative emotions experienced, disgust played a large part in the narratives to be explored in this chapter. Having looked at the cute response in the previous chapter, disgust can be seen in some ways as a polar opposite reaction to an object: cuteness attracts the subject, making them want to come closer to the cute object, to engage with and touch it, while the most prominent feature of disgust is that it repels the subject, making them recoil from the disgusting object, to get as far away as possible from it, as the narrator in the above blog tried to do, fleeing a shelter to avoid the mice living in it (the role that imagination plays in disgust is evident from the fact that the hiker had not even encountered the mice yet). The previous chapter described how psychologists Sherman and Haidt (2011: 4) write about cute entities as being "hyper-mentalized" by observers – having thoughts, beliefs, feelings, and emotions attributed to them. In the same paper, the authors talk about disgust-inducing entities as being "hypo-mentalized", meaning that the subject's ability to attribute thoughts and feelings to the disgusting entity is inhibited, and the disgusting entity (nonhuman animal or person) is more likely to be viewed as nonsentient. Thus, Sherman and Haidt describe disgust as the opposite of the cute response. This chapter will return to the idea of disgusting entities being viewed as nonsentient when discussing how hikers behaved towards animals that they found disgusting.

The function of disgust

What is disgust and where does it come from? Hikers used words like "repulsed", "freaked out", "half-ass terror", "heebie-jeebies", "gross", and "Eugh" to describe their reactions of disgust to certain animals. As mentioned previously, disgust is the emotion of being repelled away from the something (see Curtis, 2013; Durham, 2011; Herz, 2012; Lee and Ellsworth, 2013; Matchett and Davey, 1991; Strohminger, 2014). According to Durham, disgust "involves a rejection of the object" (2011: 144) and Herz (2012: 131) says that "rejection is the fundamental stance behind all that is disgusting". The experience of disgust is visceral and somatic, so much so that it can induce nausea and even vomiting (the ultimate rejection – rejecting what is already in our stomachs), bring on feelings of weakness, lower our blood pressure, and "hijack" our mind, "so that we can barely think at all" (Herz, 2012: 38). Although all perception is embodied (see Merleau-Ponty, 2014 [1945]), perhaps more than any of the other five 'universal' emotions (anger, sadness, happiness, fear, and surprise) disgust is a sensory experience, triggered by repulsive bodies, body parts, and bodily products – faeces, urine, vomit, blood, pus, sweat, saliva – processed through seeing, smelling, hearing, tasting and/or touching, and experienced in the gut. Kolnai (2004 [1929]: 44) describes disgust as an aesthetic emotion, one which focuses on features of the entity, rather than the entity as a whole. Thus, it could be argued that features such as sliminess, scaliness or a multitude of legs are what disgusts the observer, rather than the fact of being a slug, a snake or a spider.

There are a few intersecting theories around the origin and function of disgust. In "Disgust and the Anthropological Imagination", Deborah Durham (2011: 133) describes disgust as "of course, a culturally and historically specific term". This notion of disgust as stemming from the cultural is supported perhaps most famously by Mary Douglas's *Purity and Danger* (2002 [1966]). Although Douglas did not talk about disgust, per se, she suggested that dirt was "matter out of place", and therefore a cultural concept rather than a material one (2002 [1966]). According to this theory we avoid the dirty because it inserts chaos into the order that we try to create for ourselves and our environment. Supporters of the theory might argue that it is this chaos brought about by dirt – matter out of place – that elicits feelings of disgust in us.

Douglas was careful to state that her notion of dirt as matter out of place is separate to thinking about dirt in terms of pathogenicity and hygiene. In perhaps the most widely accepted theory of disgust, several researchers argue that the emotion functions as an innate mechanism for avoiding infection (Curtis, 2013; Herz, 2012; Lee and Ellsworth, 2013; Matchett and Davey, 1991). Curtis (2013) argues that pathogens and parasites caused the evolution of a variety of defence systems including impermeable skin and internal immune systems, and also including disgust: a behavioural defence system that enables humans to avoid parasites and pathogens by recognising where they might occur and instinctively recoiling from them. This is of course not to argue that all microbial life is harmful to us; indeed, Lorimer (2017a: 27) has talked recently about what he terms the "probiotic turn", which recognises the importance of certain microbes for human health, and he takes a multispecies approach to the ensuing task of "rewilding nature reserves and reworming the human biome" (2017a: 27, see also 2017b). Yet Curtis points to the multitude of infections that can be caused by body products such as faeces, nasal mucus, and blood, as well as the parasites carried by animals that people frequently react to with disgust:

> The rat carries plague [sic – the rat carries fleas, which can carry the plague] and a variety of interesting parasites such as those that cause arenaviral haemorrhagic fever, Lassa fever, and Weil's disease. Snails and

slugs carry helminthic parasites. Insects such as flies and cockroaches walk about in wastes, spreading the agents of gastro-enteric infection. Other insects, like the louse and the scabies mite, are themselves parasites. Some insects are both parasites and parasite vectors at the same time; fleas carry plague and typhus, lice carry relapsing and trench fever, ticks carry encephalitis and a variety of viral fevers. Earthworms are not dangerous to human health, but they look similar to the parasitic worms that can be found in meat and fish and that infect over a third of humanity.

(Curtis, 2013: 4)

Matchett and Davey (1991) propose that over time certain animals have become associated with the spread of disease, dirt, and contamination, and that fear of these particular animals (for example, cockroaches, slugs, snails, and worms) is evidence of the contamination avoidance theory of disgust. They cite Angyal (1941) and Rozin and Fallon (1987), who have

pointed out that nearly all disgust objects are animals or animal products, including parts of animals, animal body products or objects that have had contact with certain animals: so it would seem that avoidance of animals is a central feature of any disgust or disease-avoidance mechanism.

(Matchett and Davey, 1991: 91)

Various studies appear to support the contamination avoidance theory of disgust; for example, in their research on disgust and spider phobia in women, Mulkens et al. (1996) found that spider-phobic women showed stronger disgust sensitivity than non-spider-phobic women, and argued that spider phobia was more rooted in disgust sensitivity than in fear that the spider might harm them.

Davey (1991) suggests three ways in which animals may have come to be associated with disgust. First, by being direct sources of contamination and disease (for example, rats), second, by being contingently associated with dirt or contamination (cockroaches, spiders), and third, by possessing features which naturally elicit disgust reactions (for example, by resembling mucus, as in animals that are often perceived as slimy, like slugs and worms).

Death fear

For some researchers, disgust is not so much an innate defence mechanism, as it is an existential crisis. These commentators argue that disgust is elicited by experiences that remind us that we are animals, because remembering that we are animals means acknowledging that we will one day die (see Goldenberg et al., 2000; Herz, 2012). Terror Management Theory (Goldenberg et al., 2000) is the theory that humans have created complex cultural rituals that separate us from the rest of animal kind, in order to protect ourselves from the awareness of our own mortality. From this perspective, the sight of blood, vomit, deformed people, or a spider crawling up our arm are repellent to us because they remind us of how little control we have over our own lives and, in particular, our death. Goldenberg et al. (2000: 204) write that, "for humans, *disgust* seems to be an expression of one's disdain for or superiority to everything from foods and body products to political ideologies and immoral actions". In short, every deviation from the carefully constructed societal norms that protect us from our animality. For Herz (2012: 126), "death and disgust are fundamentally linked". This notion of disgust is not incompatible with the disease-avoidance theory of disgust; nothing reminds us more that we are going to die than the objects carrying parasites of pathogens that might kill us. However, the emphasis is a little different.

> [D]isgust is fundamentally about our awareness of our own death and our terror of it. The emotion of disgust arose from our need to protect ourselves from triggers that remind us of this truth – such as our animalistic nature – and put us in its harmful way – such as disease. Disease is a primary motivator of disgust, but it is not the psychological construct that controls it; our fear of death is.
>
> (Herz, 2012: 130)

Herz proposes, therefore, that the rejection associated with disgust is the rejection of death itself. Strohminger (2014: 482), however, argues that disgust would be a poor defence mechanism for protecting us from thoughts about death, given that disgust "seems to transfix and amplify our attention" to the very objects that supposedly remind us of our impending death.

Rather than any one theory taking precedence, it has become more common recently to acknowledge that disgust has multiple facets and functions (see Strohminger, 2014). The explanations of disgust demonstrate that, while disgust is most clearly experienced through the body – so much so that some theorists liken it more to an involuntary reaction "such as nausea, retching and the startle recoil" (Korsmeyer and Smith, 2004: 1), or a "drive" like hunger or lust (Strohminger, 2014), than an emotion – it is a complicated experience, and according to some, a "highly cognitive emotion, which provides information about features of the outer world not readily available by other means" (Korsmeyer and Smith, 2004: 2). Durham (2011: 137) describes disgust as "both non-rational, yet also a form of knowledge". It could be said that disgust is a good example of the biosocial (see Ingold, 2013), being both biological (concerned with processes of contagion, infection, and decomposition) and social (concerned with relations between individual entities), with no false division separating them.

Parasites on the trail

There are several opportunities for infection on the AT. One of the most common afflictions for hikers is Lyme disease, which can be caught from the bite of a deer tick. Lyme disease usually starts with a 'bulls-eye' skin rash and flu-like symptoms (headache, fatigue, joint pain, and swollen glands), which left untreated by antibiotics can lead to paralysis of the facial muscles, heart palpitations, shortness of breath, chronic headaches, encephalitis, and meningitis (Davis, 2012: 137–138; Logue and Logue, 2004: 167). Long-term damage from Lyme disease can include chronic arthritis and pain, and permanent damage to the heart, liver, kidneys, lungs, spleen, and eyes, and even inflammation of the brain (Davis, 2012; Logue and Logue, 2004). Deer ticks that have latched onto human skin can be hard to spot; at their smallest (and most dangerous) they are no larger than a fleck of black pepper (Davis, 2012: 137). Hikers are encouraged to check their bodies regularly for ticks. Some bloggers talked about having to end their hikes prematurely due to having contracted Lyme disease, while others talked about knowing someone who had contracted it.

Another danger on the trail is giardia, a parasite that lives in the intestines of infected animals and can be contracted by drinking water out of streams or ponds that have been tainted by the faeces of an infected carrier. Symptoms include painful stomach cramps, diarrhoea, and nausea. Hikers who do not filter their drinking water properly, or who drink directly from a stream or other water source out on the trail are at risk of being infected with the parasite.

Although Lyme disease and giardia are the most common afflictions on the trail, there are many other parasites that can be picked up by hikers, due to their immersion in the wilderness environment, as well as the difficulty of keeping good hygiene practices while living out of a backpack for six months. It is possible (if not provable) that hiker knowledge of the prevalence of parasites and their increased likelihood of picking something up while on the trail (even a sting or an insect bite can become infected and lead to complications) heightened the already inherent capacity to feel disgusted by certain entities.

Shelter mice

> When night fell, things changed. The mice declared WWIII! Typically, I have a phobia of mice. I have been forced to get past that being on the trail, but it still gives me the heebie-jeebies to have them running in circles around me while I'm laying on the floor. I would shine my light and slam my shoe on the floor while they would scamper away with little to no fear of me. I finally put in earplugs to drown out the noise. Eventually I fell asleep, but was awoken several times throughout the night to the mice trying to get in my food bag, which was hanging from the ceiling.

Hiker shelters are rudimentary constructions, usually three sided (with the fourth side open to the elements) dotted along the AT, at intervals of around a day's worth of hiking from each other. It's possible to hike the AT and almost never have to pitch a tent, camping in shelters nearly every night. Mice (mainly the house mouse, *Mus musculus*) live in and around these shelters, raiding backpacks for the food that hikers carry with them. As such, they live in a more-or-less commensal relationship with hikers – in which wild animals not only adapt to, but actually benefit from humanised spaces (see Knight, 2000: 6) – subsisting on trail mix and snack bars but not causing any serious harm. They tend only to come out at night to carry out their raids, and are known among hikers as audacious in their efforts to obtain food, chewing through anything that gets in their way, including backpacks, clothes, and containers. Blogs described them as unafraid of humans, and comfortable running over 'sleeping' hikers to get to their supplies ("a couple of mice ran over my sleeping bag and I'm pretty sure there were one or two inside my pack"). Their commensal-like existence made them anomalous in the wilderness environment of the AT and they were certainly not considered members of the 'wildlife' that hikers associated with enriching their wilderness trek.

Shelter mice were a popular subject in hiker blogs. They stole food and chewed through people's belongings, and were therefore seen as irritating and a nuisance, described as "little bastards" and "little buggers". Yet the emphasis in narratives about them was less about the destruction that they caused and more specifically on their bodies and their movement, which often evoked feelings of disgust. As seen in the earlier quote, one hiker described their "needle-ish claws" and "long bald tails", as well as the "filthy organisms that fester in their dun coats". Others also focused on their "tiny feet" and their "large" and "fat", "well-fed" and "wide bodies" as sources of revulsion. The idea of "filthy" organisms "festering" in their coats speaks to the disease-avoidance theory of disgust, with the narrator assuming that the mice must be carriers of dangerous parasites (perhaps equating them with rats), and so repelled by the idea that she "fled" the shelter and set up a tent outside.

As with many of the animals that hikers experienced disgusted reactions to, a heavy emphasis in narratives was placed on the way in which the mice moved around, their movement apparently a significant part of what gave hikers the so-called "heebie-jeebies". Mice "waddled", "pitter-pattered", "scurried", "scampered", "crawled all over", "bounded", and "charged". Their "tiny" feet went "clicking and clacking throughout the walls". Herz (2012: 81) argues that certain types of movement incline us towards disgust, including squirming, jiggling, twitching, throbbing, and writhing, which are apparently likely to correspond to "sick and contaminating things", and seem to have in common that they are unpredictable and irregular movements, much like those of the shelter mice. Most of the time mouse raids happened in the dark, while the hiker was trying to go to sleep, or had been woken from their sleep, and without being able to see the mice the effect of hearing them moving around the shelter was amplified. Hikers also described the sounds of "chewing", "nibbling", and "gnawing" without being able to tell what the mice were chewing on, and whether it belonged to them.

Some mice were particularly fearless in their contact with hikers, running up and down their sleeping bags, crawling on and around them. In one narrative,

> the mice last night were a little bit out of control. I woke up at one point, with the hairs on the back of my neck prickling, to see the largest, fattest mouse I've ever laid eyes on sitting inches away from my nose, staring quizzically at me.

Another described lying on her stomach with her arms by her side and feeling a mouse nibbling on her pinkie finger, after which she kept her hands in her sleeping bag for the rest of the night. One blogger posted: "Oh, and a mouse ran right over my mouth while I was reading in the shelter. Eugh".

Mouse activity in and around shelters had differing effects on the hikers camping there. Most talked about having trouble going to sleep, being woken by the noise the mice were making, or staying awake all night listening to them. One tongue-in-cheek comment talked about falling asleep to the "lullaby of mice scattering around us". One hiker, in anticipation of being disturbed by mice on her first night in a shelter, said that she was thinking about taking a sleeping pill that night. Another described watching "their wide bodies waddle up and down our tent one night at Long Branch Shelter, whilst clutching our head torches in half-ass terror, fearful they'd chow through our tent...". A few hikers were so upset by the mice that they left the shelter and pitched a tent outside, or packed up very early in the morning and set off hiking. Others talked about trying to scare the mice by moving around the shelter, slamming their boots on the floor, and trying to "shoo" them away.

One lengthy narrative offered a moment-by-moment account of what happened when a mouse entered the tent that the blogger and her hiking partner were camping in. On realising the mouse was inside the hiker described thinking "oh sweet heavens please no!" and then cinching her sleeping bag as tightly as it would go around her face, huddling inside it. When the mouse ran down her sleeping bag the blogger responded "Oh. My. Word.... A mouse just scurried down your body and you were conscious for every single moment". She describes going into a "daze of total panic and hysteria" and beginning to "hyperventilate violently". Eventually she was able to calm down enough to pass a pair of gloves to her hiking partner, who put them on, picked up the mouse and threw him out of the tent.

> Like any good war story, the audience should probably know where we are now. CC and I awoke the next morning and were able to digest the madness of what happened. It's to the point where we can laugh openly about it now, which is good. I also did apologise for being the most worthless girl ever, which led CC and I to our first ever debate: which is worse? Snakes or mice? We've come to a consensus where CC will chuck the mice and I will handle the snakes. Let's just hope it never comes to either again.

Disgust, fear, and the perception of shelter mice

Although many of the narratives about shelter mice portray them as irritating and somewhat disgusting, the above story illustrates the fact that for some people mice were both disgusting and frightening, causing them to experience "terror", "panic", and "hysteria". For many commentators, fear and disgust are closely related emotions (see Herz, 2012; Kolnai, 2004 [1929]; Lee and Ellsworth, 2013). Lee and Ellsworth (2013: 274) write that, "in the absence of perceived agency and justification, the action tendencies of avoidance and withdrawal, and the subjective experience of weakness and dependence, physical disgust resembles fear", pointing to the fact that reactions of both fear and disgust involve the experience of a lack of agency in, or control over, an encounter with a disturbing entity. Hikers who lay in the dark wondering feverishly whether the chewing they could hear was the destruction of their belongings can certainly be said to have been feeling the effects of a lack of agency in the presence of shelter mice.

Herz (2012) describes disgust as a type of fear. Where one type of fear – such as that of immediate dangers like being chased by a tiger – is urgent, disgust is "an unfolding and cognitive emotion; it protects us from creeping dangers that we have to figure out, dangers that are slow in their deadliness, and of which disease, contamination and decomposition are the foremost threats" (2012: 80). To illustrate the connection between disgust and fear Herz points to phobias of blood, vermin, and insects, all of which do not usually inflict direct fatalities, but whose "ability for destruction comes through the gradual and uncertain path of infection" (2012: 80). Matchett and Davey's study (1991) on the disease-avoidance model of animal phobias found that sensitivity to disgust and contamination was directly related to how afraid people were of certain animals (rats, spiders, cockroaches, maggots, snails, slugs) that were not normally considered as animals who might attack and harm humans.

In their introduction to Kolnai's *On Disgust*, however, Korsmeyer and Smith (2004: 19) point out that a subject experiencing fear has an elevated pulse, while disgust actually slows the heart rate. Also, the two emotions are processed in different parts of the brain: the amygdala for fear and the insula and basal ganglia for disgust. In his phenomenological exploration of disgust, Kolnai (2004 [1929]) also points to a few essential differences between disgust and fear. He views the person experiencing fear as disinterested in any particular quality of the threatening object, other than its ability to harm. In contrast, as mentioned previously, Kolnai views disgust as disinterested in the entity as a whole being, and entirely focused on the entity's particular disgusting characteristics – essentially the opposite of how we perceive the frightening entity. Those qualities here included tiny feet, wide bodies, and fur that might contain parasites. He also points to a difference in status between the object that disgusts us and the object that frightens us.

> For whereas fear intends its object as something threatening, as something 'stronger than myself' (even in those cases where I feel that I can,

if necessary, repel the attack, even overpower the attacker), there is in an intention of disgust a certain low evaluation of its object, a feeling of superiority. What is disgusting is in principle not threatening, but rather *disturbing…*

(Kolnai, 2004 [1929]: 42)

Although Kolnai describes the disgusting entity as not being physically threatening, his description of it as "disturbing" implies that the entity represents a kind of psychic threat to the observer, who may feel deeply unsettled by certain qualities inherent in the entity, one of which is deemed to be its supposed inferiority to the observer. I am not sure that this assertion is entirely correct when considering the relationship that exists between hiker and shelter mice. Although the mice were significantly smaller than the human hikers, and therefore supposedly vulnerable to them, it was frequently hikers who fled from mice, and not the other way around. It is perhaps possible to hypothesise that shelter mice were disturbing to hikers precisely because the seemingly obvious relationship of human dominance and mouse infer-iority was simply not recognised by the mice, who asserted their own con-trol over encounters by touching hikers when they wanted, and taking from them what they wanted. Indeed, one blogger even spoke approvingly of her hiking partner, who had cleverly made sure that none of their belongings were destroyed, or food taken. "Nate left a hefty pile of sunflower seeds as a 'peace offering'…and it worked!" The fact that hikers felt compelled to leave an "offering" for mice speaks to a relationship in which the mice very much dictated the terms of encounters.

Interestingly, Kolnai's interpretation of disgust gives it something in common with the cute response, namely, that the subject of both disgust and the cute response is supposedly regarded as inferior to the observer. In fact, shelter mice have many of the qualities that usually elicit a 'cute response'. They are small, furry, chubby looking, vulnerable to harm from humans, but confident in approaching them. Yet almost none of the narratives reviewed described them as cute or appealing in any way (although one did describe a hiker building a "nest" for newborn mice). When ponies or even chipmunks hassled hikers for food they were considered "adorable", but mice approaching hikers for food were at best a nuisance, and at worst revolting and panic-inducing. Why were hikers so uncomfortable around mice?

Context, ambiguity, and control

Firstly, it's possible that shelter mice were equated with rats in the minds of some people, the two species being phylogenetically similar, and both being generally classified as 'vermin'. Arluke and Sanders (1996: 178) describe vermin as being "lower than freaks", who "stray from their place, cross human-drawn boundaries, and threaten to contaminate individuals or the environment". As we have seen, disgust is an emotion elicited specifically by the perceived threat of contamination, whether real or imagined. Yet the idea of vermin, like so many animal categories, can be seen to be socially constructed. For example, Arluke and Sanders (1996) point out the fact that mice in a biomedical research facility do not elicit the same disturbed feelings – although the mice in the biomedical facility would presumably be contained rather than free, and be perceived as 'clean' rather than 'infested'. Indeed, Herz (2012: 35) argues that a large part of what disgusts us depends on how we "meet" the offending object, using saliva as an example: our feelings about the saliva in our mouth vs. drinking our own saliva from a glass. Meeting mice in a laboratory may not have been disgusting for the bloggers who expressed repulsion about shelter mice, but meeting them in trail shelters was. In another example of the constructed nature of the category 'vermin', Song (2001) writes about the annual pigeon shoot that takes place in the town of Hegins, Pennsylvania. The shoot originally started as a way for poor people to obtain pigeon meat, but the pigeons are now viewed as inedible pests, a status that seems to have been ascribed at around the same time that the town began importing pigeons from nearby cities for the shoot. Thus, shelter mice may be responded to as repulsive due to the perception of them as vermin/pests, a status which in itself is questionable and socially constituted.

Secondly, one could argue that their ambiguous status is in part responsible for the perception of them. They live along the AT, but not as part of the local 'wildlife'. Instead, they house themselves largely in and around shelters, constructions built exclusively to protect human hikers from the elements, and which can therefore be considered to be 'domestic' domains. The mice's inhabitancy of shelters and reliance on hiker food makes them not 'purely' wild, but they are not domesticated either. In *Purity and Danger*, Douglas (2002 [1966]: xi) calls out the "cognitive discomfort caused by ambiguity", stating that "ambiguous things can seem very threatening". The mice's ambiguity is easily equated with *impurity*, which is more likely to make them disgusting.

Thirdly, they were seen as invading human boundaries, by dwelling in and around human shelters, by crawling inside backpacks and sleeping bags, and by running over hikers, and even nibbling their fingers. Jerolmack (2008: 87) points out that there are numerous ways in which animals can be ranked and evaluated, and one of the most crucial of these is the "spatial dimension", with animals' "unchecked presence" symbolic of disorder and impurity. Whereas the cute object waits or appeals to be touched by the human subject, shelter mice touched hikers (they are the touch-ers, not the touch-ees, so to speak) as they went about their business of finding food. The mice asserted their agency in touching hikers and in taking the food they wanted (rather than making an appeal for food), and hikers felt an associated lack of agency in not being able to stop the mice from touching them or from taking their food. If they wanted to sleep in a shelter they had no choice but to tolerate being walked over, investigated, and probed for munchies.

Candea and da Col (2012: 13) describe the concept of hospitality as riddled with tensions and ambivalence, not least of all when thinking about "the parasite", seen as "one having the ability to cross ontological boundaries and confusing the subjectivities of host and guest". Indeed, when thinking in terms of hospitality and spatial sovereignty, it is difficult to identify with any certainty who the "host", the "guest" or the "parasite" was in the shelter mice–hiker dynamic. The mouse, who lives in a shelter all year around, will surely see the shelter as his home, and the hiker as a (beneficial) visitor to it. The hiker, who views the shelter as having been built for her – and other humans' – use, perceived the mouse as an intruder in her domain, despite the mouse's permanent habitation there. In his article on "The Unwelcome Crows" van Dooren (2016: 197) describes the act of welcoming or excluding another as an act of appropriation, as claiming "the right to decide who comes and who goes". For van Dooren, the Anthropocene is defined by an approach to the world that makes humans view themselves as "the hosts, and others, permanently, guests in our space, by our grace" (2016: 201). Thus, it is possible to see why hikers were so perturbed by the presence of mice in shelters; mice were perceived as unwelcome guests, taking advantage of unwilling hosts, rather than as residents in their own homes. Returning to Kolnai's (2004 [1929]) description of the disgusting entity as disturbing rather than threatening, it is possible to describe hikers as disturbed by shelter mice because they took what they wanted from hikers (rather than endearingly requesting or charmingly demanding it, like cute entities did), clearly viewing the hiker's resources – both food and the shelter itself – as their own, thereby disrupting the common assumption of humans as the decision-making hosts of the world. In arguing for an approach that recognises our cohabitation of the world with other species, van Dooren therefore describes hospitality as an unsuitable basis for "responding to others in *multispecies* worlds" (2016: 204, emphasis in original), because anyone who views themselves as a host is inevitably staking a claim on a space. Yet animals who take what humans view as 'belonging' to them inevitably come to be viewed as pests, as Marvin (2001) notes in his writing on foxhunting, with the fox being viewed as a pest precisely because of his predation on domestic animals such as sheep and chickens, who are considered to 'belong' to human farmers, rather than predation on wild animals, such as rabbits. Shelter mice viewed hiker food as 'fair game', and could thus be perceived as either very rude guests or very bad hosts; and transgressive in either instance.

Yet although hiker narratives focused on the physicality of shelter mice, the way they looked, and in particular, the way that they moved, it wasn't these qualities *per se* that made them so disturbing to hikers. Rather, the significance of their scurrying, pitter-pattering movements was that the mice were *unpredictable* in where they would go, and what they would do next, and this unpredictability was a significant factor in how mice established *control* over encounters between themselves and human hikers. Therefore, it could be argued that it was the assumption of dominance by these small, supposedly vulnerable mammals, and hikers' inability to defend themselves from the onslaught of probing claws and nibbling teeth, that resulted in feelings of fear or disgust.

Finally, shelter mice usually come as a multitude, and multitudes of disconcerting animals are far more disgusting than single animals.

Multitudes

> It wasn't even dark yet before the mice started to gather, in Sparrow's words, to quote her exactly, "the mice are gathering"… [a] hoard of mice, collecting themselves in droves, I told myself, inside the shelter…

Certain animals on the trail seemed all the more disgusting when they appeared in multiple numbers, which amplified their effect on hikers. These animals included shelter mice, as well as worms, leeches, slugs, caterpillars, spiders, and snakes. One hiker described "so many leeches in the water. Little black ribbons of nightmares effortlessly slipping between rocks and twisting around each other…", and later described her path being blocked by multiple snakes in an "Indiana Jones nightmare". Several hikers commented on how many caterpillars there were in the woods, with one referring to "the great caterpocalypse" and having to pull dozens off his body every day, and others talking about how there were so many caterpillars suspended from tree branches that caterpillar poo "fell like rain". Others talked about how on certain places on the trail there were so many spider webs across the path that they had to take turns with their hiking partners in leading the way and parting the webs with their trekking poles ("annoying webs that were strung out across the trail every single day and stuck to us like glue!"). One hiker accidentally stuck her trekking pole in a wasp's nest, and she and her partner were stung multiple times by the wasps that swarmed out of it. Caterpillars were referred to as an "incalculable legion", and people talked about being "plagued" by insects, including "swarms" of mosquitoes. One hiker talked about her dog's discovery of worms along the trail.

> As we were hiking through the beautiful wilderness, I noticed that Marley was creepily hunched over. I asked him, what are you doing? He told me he wanted me to meet my nemesis. I knew it was a worm, so I cautiously trekked over to Marley. But as I started hiking again, I saw one close by. Then another. And a gross swollen one writhing miserably in the mud. The few seconds it took me to realize that I was surrounded by hundreds of earthworms felt like hours. I felt like I was in the middle of a particularly bad nightmare as I sprinted down the rest of the mountain, noticing every worm as I passed by.

Herz (2012: 103) writes that swarms are often elicitors of disgust, using ants as an example; two ants running across the kitchen floor are tolerable, but a swarm of them "scuttling across your kitchen floor is repulsive".

> The swarm is vibrant with life, but it is also overwhelming, disorderly, chaotic and uncontrolled. The swarm has more potential to infect us than

the few, because the more there are of any living thing out there, the more chance there is that they will get onto you or into you, and...a primary objective of disgust is to keep the outside away from our inside.

(Herz, 2012: 103)

As the hikers who were stung by wasps, those who had to pull caterpillars off their bodies and belongings, including the lip of their drinking bottles, or pull ticks out of their beards, the presence of multiple animals made it harder to keep the 'outside away from their inside'.

Kolnai (2004 [1929]: 55) locates the overwhelming effect of a plurality of "disgusting" animals with what he calls "an indecent surplus of life... an abundance that, true to nature, points once more to death and towards putrefaction...", indicating that he, too, sees disgust as the human response to reminders that we are mortal animals. Like Herz, he sees the threat of the swarm in the greater possibility of it infecting the human subject, arguing that this surplus life "endeavours to break through any boundaries which may be set upon it and to permeate its surroundings" (2004: 73).

What is real is a frequent preference for putrescent organic material; what is apparent...is the impression that they themselves are somehow part of the stuff, as if they had originated from it, as if their frantic teeming activity were a phenomenon of life in decay. Altogether it is in general the strange coldness, the restless, nervous, squirming, twitching vitality which they exhibit...

(Kolnai, 2004: 58)

The presence of a swarm of animals also makes it harder to focus on any one animal, thereby potentially increasing the likelihood of them being seen as an indistinguishable mass, rather than a collection of individuals. Edelman (2005: 126) writes that in Victorian times rats were reviled in part because "they were mass creatures as opposed to individuals". When thinking about disgusting animals being more likely to be viewed as "non-sentient" as Sherman and Haidt do (2011), a multiplicity of already 'disgusting' animals is therefore even more likely to be seen as nonsentient than an individual 'disgusting' animal.

The practical advantage to animals of being encountered in groups or swarms was that hikers were more likely to run away, otherwise actively avoid them (as some people did with mice, worms, leeches), or learn to tolerate them (as people did with spiders, caterpillars, and other insects) than to try to kill or otherwise eliminate them, or force them to leave the vicinity, as they frequently did when in uncomfortable confrontations with an individual animal.

Confrontation with a skunk

Although skunks did not feature often in hiker blogs, one person wrote two lengthy narratives around a confrontation that she and her hiking partner had with a skunk (*Mephitidae*). The skunk approached their tent one morning, and seemed to be trying to get in by crawling underneath it, which made their air mattresses bounce up and down.

> I was totally freaked out...my heart was racing and my hands were shaking. The skunk was either totally acclimated to people, or rabid! It would not back off... Not long after we had irritated the skunk enough to make it leave the bottom of the tent it showed up on Marc's side of the tent again, and tried clawing and biting at the screen to get in! At this point, Marc punched him in the face – and still he came back! He punched him again! OMG! I felt like I was starring in a late night horror movie!

When the skunk backed away the two hikers quickly packed up their gear, intending to "get the hell out of dodge".

> When I got out of the tent, Marc noticed that I had forgotten to zip the bottom of my side of the tent!! Cujo was about to get in! No more Mr Nice Guy! Marc picked up a stick and started whacking that deranged beast on the head! After a good deal of whacking the skunk backed off, but not quickly and not without creating a stink.

The skunk did not spray the hiker directly, but did spray into the air, after which the two hikers described themselves running down the trail, "leaving our hiking companions to deal with the stench and a rabid, deranged skunk!" They later discovered that the smell had permeated their clothing, and had to go into a town to buy vinegar, baking soda, and detergent, which they had found out would help to get the smell out of their clothes.

> As we were ordering our lunch at the DQ, a kid behind the counter said "Ewww, I smell a skunk!" (so much for thinking we had untainted clothes on)... I felt like a social pariah!

People who had disgust reactions to shelter mice and other animals described their characteristics in detail, including their aesthetic features and the way that they moved. In line with Kolnai's (2004 [1929]) description of fear having no interest in the characteristics of the frightening entity, this blogger did not discuss how the skunk looked or how he moved his body, focusing only on his perceived threatening actions towards her and her partner. The skunk was not viewed as disgusting, but as an immediate threat to their physical safety.

The reaction of the blogger's partner to this perceived threat – repeatedly punching the skunk and hitting him with a stick – seems like quite an extreme fear response to the situation. There was no mention in the narrative of the humans involved in the incident having considered that their tent might be pitched somewhere significant in the skunk's territory. We also don't find out what happened to the skunk afterwards, although the blogger describes seeing him heading towards another hiker's tent as they left the campsite. Having subsequently looked up the symptoms of rabies online, she theorises that he was afflicted with it (which she mentioned suspecting during the incident). The fact that he was viewed as "rabid" and a "deranged beast" seemed to make his actions more frightening, yet the potential of being infected with rabies interestingly did not appear to make him disgusting in the eyes of the hiker – perhaps because of prioritising the immediate threat of being bitten or clawed over the more slow-moving threat of infection.

Instead of the possibility of being bitten and contracting rabies, it was being sprayed by the skunk that transformed fear of him into disgust. Several theorists, including Kolnai (2004 [1929]), describe smell as the sense most attuned to inducing disgust (ahead of sight, touch, taste, and with sound last). The skunk's 'disgusting' odour was transferred onto the hikers, who became 'contaminated' with it, and who then themselves became disgusting to the people who came near them, much to their dismay.

Snakes on a trail

> I was pushing myself to get to the first hut before dark…and I was already stepping over the snake when it started rattling with what felt like an explosion of sound. I didn't consciously recognise the sound at first, in fact I originally thought it was simply the cicadas I'd been hearing for a while already, just echoing oddly close. Despite this, I still felt uncomfortably scared, with fear pulsing through my body at high speeds. I jumped in the air and leapt forward several steps. It was only when I looked back and saw the shaking tail raised in the air that I realised what happened.

Several species of snake inhabit the land that the AT crosses through. The two poisonous snakes that live along the trail, and were most frequently featured in hiker blogs, are the copperhead (*Agkistrodon contortrix*) and the timber rattlesnake (*Crotalus horridus*). Both snakes have become rarer along the trail, due in part to habitat loss, but also to "indiscriminate killing" by humans (Adkins, 2000: 184).

Hiker narratives reflected strong feelings about snakes on the trail. Like bears, they were often viewed as charismatic and emblematic of a life lived in the wild. They were undoubtedly the most viscerally feared animal on the trail, but also one of the most alluring. As one blogger wrote,

> we were approached by a…hiker who warned us that he had just run into a pretty pissed off rattlesnake…naturally I grabbed my camera with the mission of finding it…I was pretty freaked out walking along the trail trying to find the snake.

Photographs of snakes were frequently posted on hiker blogs, perhaps partly in order to demonstrate the willingness of the photographer to approach the snake. Several of the narratives about snakes expressed fear and awe ("I had a terrifying surprise"), while some attempted a humorous, self-deprecating tone when describing their snake encounters, similar to the self-deprecating humour used in some bear conflict narratives ("ok, death by rattlesnake is just for dramatic effect...I was reminded that I am a guest out here. The animals and mother nature rule shit"). It is also clear from a few blogs that, for a minority of hikers, it was considered acceptable to kill, and sometimes eat, snakes that they came across.

Where mice were disturbing in part because of their diminutiveness, being "little buggers" with "tiny feet", snakes were also most often described in terms of their impressive size, as "long", "huge", "huge and terrifying", "4ft" or "5ft", and one rattlesnake "as thick as Lydia's arm". As with mice, and other discomfiting animals, the way that they moved their bodies was emphasised in narratives about them; they "wriggled" and frequently "slithered", or lay "coiled up" and "camouflaged perfectly". Many bloggers wrote about how they didn't see the snake until they were almost stepping on him, which proved a large part of their sudden terror when they realised their unexpectedly close proximity to a snake. People experienced fearful reactions to snakes despite the fact that most blogs described them as either lying on the trail or in brush, or moving to get away from the hiker, with only two describing a snake that "slithered" or "lunged" towards a hiker. Rattlesnakes characteristically stood their ground and rattled their tales at hikers who came near to them, and in one instance an individual was described as having reared his head at a hiker, although "it did decide to move along quickly" afterwards. Despite the lack of any overtly threatening action by snakes towards humans, most hikers still interpreted snake presence as highly threatening to their safety. Many chose avoidance, and talked about stepping back from them, skirting around snakes when they were lying in the path, and generally getting as far away from them as possible.

> About two thirds of the way down I spotted this critter, two steps before I would have had a close encounter. Somehow I think I backed up 20 feet without touching the ground. I hike without my hearing aids, so I couldn't hear the warning it gave. But I saw the rattlers vibrating...it took me 5 minutes or so to get up the nerve to slowly step around it with about a 5 foot space. I knew that Falcon and Poncho were 10 minutes behind me, and that one of them hiked without their hearing aids...so I waited for them. Sure enough they didn't see the snake until I pointed it out when they were 10 feet away. We agreed, no way should you have to battle snow and ice in the morning, and dodge rattlesnakes in the afternoon. That is our shared catastrophe. Cold weather and a rattlesnake. Whenever I run into these two on the trail they tell others that I 'saved them' from a rattlesnake. Wish I'd gotten their picture with the snake. But we were too rattled.

Some hikers, however, seem to have felt compelled to approach, and even interact with, snakes. Indeed, according to Logue and Logue (2004: 158), ankle and leg bites are very rare on the trail, and the majority of snakebites to hikers occur on the hand, because hikers have picked up the snake (venomous or otherwise). In Klein's ethnography of the AT (2015), she quotes her informant "TV Idol":

> If I see a spider, I'll flick it off. I won't crush it because then I'll probably get all its nasty shit all over me. Yeah, snakes, if it's a garter snake, maybe we'll poke it a little bit, get it a little pissed off. I mean, you get bored out here.
>
> (Klein, 2015: 143)

Interactions written about by hiker-bloggers included approaching to watch or take photographs of a snake (including seeking out a snake another hiker had mentioned in order to take pictures), running at snakes while yelling or clicking hiking poles (to try to get them to move off of the trail), pulling a snake out of a fire pit that he was trying to burrow into (seemingly to get away from interested hikers), and grabbing a snake and picking him up, either to get a closer look, take a picture holding the snake, or apparently occasionally so that a hiker could kill the snake.

> We met Serpent and Grinner on top of a mountain just as they were getting ready to leave. I pointed out a snake to Hope, and Serpent immediately jumps into the brush and grabs the snake with his bare hands. This wasn't his first snake. In fact, he got his trail name because he was catching so many serpents.

As with bear encounter humour, some of the discomfort around snakes was translated into wry commentary on the situations encountered, like the blogger who wrote that he and his companions were too "rattled" to get a picture with the rattlesnake he'd come across. When a hiker found out that a family of copperhead snakes was living underneath the shelter he planned to stay in (copperheads are often encountered in groups, being fond of the company of their conspecifics), he commented that "on the plus side, the snakes meant no mice". Another blogger interpreted a rattlesnake rattling at him as "their lazy way of saying, 'Bro, please don't step on me'", although admitting that "it's not so much the act of being bitten as it is the cascading effect of seeing one". One blogger walked past a large black snake that "tried to convince us it was a rattlesnake by shaking its tail against dry leaves", giving the snake credit for "a fantastic effort".

> The snakes were all the same kind. I stepped on one and sat on one. Owen got in an argument with one, and named another "Karlor, keeper of the privy"... the one I sat on was sunning itself on a rock ledge next to a logging road where we decided to take a break. I used the rock as a seat to eat a snack and re-tie my shoe. When I got up the snake was back on the rock ledge, just chilling and eyeing me when I turned around. Apparently

he was extremely comfortable with the warm rock sheltered by my warm butt. Can't say I was the same... Karlor just liked to sun himself on the rock step to the privy... every time you approached the privy he would slither under the rocks. As soon as you left, he would be back guarding the door.

The fear and fascination of snakes

Much has been written about the supposedly innate human fear response to snakes (LoBue and DeLoache, 2008; Matchett and Davey, 1991; Mundkur, 1994; Öhman and Mineka, 2003; Wilson, 1984). In *Biophilia* (1984) Wilson refers to nonhuman primates who exhibit a fear of snakes, including guenons and vervets who, on seeing snakes, emit a call that warns other members of their group about the presence of the snake, and rhesus macaques, who will back away, stare, shield their faces, bark, and screech on sighting a snake. Furthermore, Wilson states that monkeys raised in a laboratory with no prior experience of snakes show similar responses to those in the wild, implying that the fear response to snakes is at least partially instinctive. Mundkur (1994) looks at the reactions of human children, whom, she argues, are emotionally predisposed to experience wariness of a few specific animal species, most markedly the serpent. According to Mundkur, the chronological development of ophidiophobia – the fear of snakes – is remarkably similar in chimpanzees and human children. Furthermore, she argues that while the neurotic fear of practically all other animals tends to decline among older children, the fear of snakes "shows a precisely opposite ontogenetic tendency" (1994: 157). In their study on attention to fear-relevant stimuli by adults and children, LoBue and DeLoache (2008) used visual detection tasks to show that both children and adults can detect snakes more rapidly than nonthreatening stimuli (they used flowers, frogs, and caterpillars). They argue that their results provide evidence of "evolutionarily relevant threat stimuli in young children" (2008: 284).

Although our fear of snakes is argued to be innate, we are apparently not born afraid of them (see LoBue and DeLoache, 2008; Öhman and Mineka, 2003; Wilson, 1984). Instead, it has been proposed that humans are predisposed to learn our fear of snakes from an early age; an example of what LoBue and DeLoache refer to as "prepared learning" (2008: 284). According to Öhman and Mineka (2003), humans and monkeys learn to be afraid of snakes more easily than almost any other stimuli, either through direct or vicarious conditioning. They state that snake phobia can be activated simply by seeing pictures of snakes, while Wilson (1984) uses the examples of a toy snake being thrown at a child, or a scary story about a snake told around a campfire, as examples of snake-fear activators in human children. The "preparedness hypothesis" (see Mulkens et al., 1996), which proposes that fear reactions to stimuli which once posed a threat to our ancestors are quickly learned and difficult to un-learn, accounts both for disgust responses to supposedly 'contaminated' animals, as well as to snake fear, with researchers arguing that the quickly acquired fear response to snakes is most probably linked to the danger that their reptile ancestors posed to our human ancestors (LoBue and DeLoache, 2008; Öhman and Mineka, 2003; Wilson, 1984). Öhman and Mineka (2003) argue that fear of snakes is fairly independent of conscious cognition, and that people who encounter snakes in the wild sometimes freeze in fear before even realising that they were about to step on a snake, as evidenced by the blogger who talked about fear "pulsing" through his body before he had recognised the rattling sound he heard as belonging to a snake.

LoBue and DeLoache (2008: 288–289) consider certain physical characteristics in snakes as being pivotal in drawing human attention to them. The first is the way that they move – by slithering – which they obtained evidence for in a study on human infants' responses to snakes. The second and third are attributes that distinguish snakes from any other animal, namely, their long, limbless bodies, and their consequent ability to coil themselves up. LoBue and DeLoache's argument is supported by hiker narratives, in which hikers focused repeatedly on how long the snakes they saw were, their slithering motion, and how they were often found coiled up (which was considered particularly frightening because their coiled-up position meant that they were often not spotted until the last moment).

LoBue and DeLoache cite Isbell (2006), who published an analysis of the origin of the primate (including human) visual system, in which she proposed that some of the basic properties of our vision evolved specifically in order to facilitate the detection of snakes, which speaks to just how ancient our sensitivity to snakes may be.

Fascination

Snakes were described by hikers as being frightening far more than as disgusting. The focus on their bodies and movement is, according to Kolnai (2004 [1929]), a characteristic of disgust and not fear, yet here this focus, rather than indicating disgust, appears to be indicative of a kind of horrified fascination that many hikers experienced when in the presence of snakes. Indeed, as well as feeling repelled by the sight of a snake, hikers behaved in ways that suggested an attraction to snakes. Kolnai (2004 [1929]) describes the dangerous or threatening entity as only interesting to us in so far as how much of a threat it poses, which stops us from experiencing any interest in its features or characteristics. Yet Wilson (1984: 86) writes that "the mind is primed to react emotionally to the sight of snakes, not just to fear them but to be aroused and absorbed in their details". Herz (2012: 131) writes that we are "perversely attracted to and lured by death, destruction, and that which disgusts us most" (one hiker described dozens of leeches twisting around in water as "horribly beautiful"). The combined fear of and fascination with snakes may have had something to do with the killing and consumption of them by some hikers on the AT, which will be discussed a little further on.

As talked about in Chapter 2, Beeman (1999) describes the basis of humour as incongruity, and some bloggers used the general incongruity of living with and among snakes as a source of comedy in their narratives. This includes the incongruity of not being able to hear a rattlesnake's warning because of deciding not to wear your hearing aid, of sharing a toilet outhouse with a snake, and of attempting to find comfort in a family of snakes living beneath your shelter, because at least they will keep the mice away. The self-deprecating humour employed was part of a counter-narrative occasionally used by hikers (and seen in narratives about black bear encounters as well) of being inherently *un*prepared and *un*equipped for living in the wild, and yet learning to get along in it anyway. Unlike the hikers on the trail, snakes can be seen as perfectly at home in the AT environment; they are able to camouflage themselves in the leaves or against rocks, rattlesnakes are able to warn others away (while other species of snake are able to imitate their rattle by shaking their tails in dry leaves), they are able to sit still and 'taste' the air for their prey, hunt down their food, and poison attackers or prey. Their bodies are able to take them wherever they need to go; they can go up trees, disappear into tiny crevices or vanish into the undergrowth. The fact that they appear perfectly adapted for a life in that environment, that they are more at home there than human 'visitors' can possibly be, must surely be part of their fascination for hikers. Their presence can serve as a reminder to hikers that humans are not the 'hosts' of the world, and that on the trail, hikers are reliant on the hospitality of the snakes there for their continued safety.

Killing animals on the trail

Blog posts that talked about killing animals on the AT were very rare, but out of ten posts found that included descriptions of animal killing, five involved the killing of a snake or snakes. Two of these involved what were described as accidental killings, both at fire pits.

> We made camp...at one point Looper jumped back and said, "there's a snake!" pointing to a small brown snake beside the fire ring. Since it was near our game we tried to pull it out, but it burrowed in the ring. I lifted the rocks, revealed it. But as Looper lifted it up with hiking poles the poor snake slipped into the fire. It writhed and spasmed on the coals for seconds, then stiffened: the image is burned in my mind. We later realised it was a harmless worm snake and turned it into a joke. Because not only did Looper 'accidentally' kill this snake, her confession at the priest had been about killing another one!

Another story described the killing of two juvenile water moccasins (*Agkistrodon piscivorus*), not by hikers but by canoers, who showed the hiker-blogger the recently deceased bodies of the snakes. "Apparently these snakes can be highly venomous so I walked up to the closest campsite to warn a family".

The other two narratives about snake killing both talked about the killing and eating of a rattlesnake. In one, the author of the blog met a hiker called Skinner, who got her trail name when her boyfriend killed a rattlesnake and she skinned and prepared the dead snake for them to eat. The other included a more detailed narrative.

> I was hiking ahead of my group and came upon a rattlesnake laying across the trail. I immediately held my dog back, told Coo (who is terrified of snakes) to stay behind, and told Golden (a natural survivalist) to come ahead. Golden immediately got excited because he had already mentioned wanting to eat rattlesnake, and before he even saw it assumed I found one. He got his wish and the rattlesnake met his ending. I will forever have nightmares about this process, and will not go into detail… but, that day, I learned how to decapitate, skin, and degut a rattlesnake without even wanting to know…I refused to eat it and my dog wouldn't even try it, but everyone else enjoyed it…
>
> [Editor's note: Golden's behaviour was a clear violation of Leave No Trace, very sad, and something we condemn. This story will remain on this site as an example for others to learn how to absolutely not behave… this is something that is happening on the trail, and it should be addressed as opposed to ignored. For those who need a reminder on the principles of LNT, visit their website, and check out one of the many articles we've posted on the topic].

The notion that snakes are 'killable' (that not only is killing snakes not wrong but might actually be a good thing to do) comes across both in the high proportion of animal-killing stories that centred around snakes, and even more so, in the (very rare) editor's note at the end of the previous blog, which mentions the killing of snakes as "something which is happening on the trail". Indeed, the overall very small number of posts about snake killing on the trail (and animal killing generally) cannot be assumed to be representative of how much the killing of snakes and other animals is going on, given that potentially not every hiker would be comfortable blogging about either having participated in or observed a killing, on a public forum like the Trek.

What is it about snakes that made them – at least for some people – inherently killable? The perceived danger posed by two water moccasins meant that a couple of canoers felt justified in killing them (the fascination associated with snakes then demonstrated in the way that the canoers spent some time showing their dead bodies to passing hikers). Their extreme otherness surely plays a part as well, including those features that LoBue and DeLoache (2008) mention: their slithering movement, long, limbless bodies, and consequent ability to coil themselves up. This radical otherness in itself (regardless of the physical threat posed by the snake's venom – or lack of it) can be interpreted as a threat to be eliminated, or at the very least make it less likely for hikers to ascribe thoughts and feelings to the animal. The fact that snakes are interesting to people proved to be a danger for them, as evidenced in the narrative about a burrowing snake being pulled out of the ground on a hiking pole, only to slip into a fire. It also proved dangerous because it was the lure of eating rattlesnake meat that caused the death of the two rattlers.

As with black bears on the trail (see Chapter 2), snakes were clearly used by some people for the purposes of edgework – the practice of engaging in deliberately high-risk behaviours (see Lyng, 1990) – specifically those hikers who on sighting the snake grabbed and picked him up. Rattlesnakes, which can hurt and even kill a human being with a single venomous bite, containing various hemotoxins (for attacking blood cells) and neurotoxins (for attacking the nervous system), are indeed the perfect props for this kind of voluntary risk taking, which borders on the anarchic.

> [E]dgeworkers claim to possess a special ability…this unique skill, which applies to all types of edgework, is the ability to maintain control over a situation that verges on complete chaos, a situation most people would regard as entirely uncontrollable. The more specific aptitudes required for this type of competence involve the ability to avoid being paralysed by fear and the capacity to focus one's attention and actions on what is most crucial for survival.
>
> (Lyng, 1990: 859)

Lyng's description of edgework situations as verging on complete chaos is resonant of the distinction that Marvin (2006: 24–25) makes between domestic killing and wild killing, in which domestic killing (such as that of farm animals or the euthanasia of companion animals) is described as being orderly and inevitable, while wild killing (such as hunting, or in this case, impromptu rattlesnake killing) is "disorderly and certainly not inevitable, because it is based on the lack of continuous control of wild animals by humans". Once control has been achieved through the death of the animal, however, "the hunted animal belongs fully to the hunter, and being recognized as ultimately and individually responsible for the death of that particular animal is a significant constitutive feature of the relationship between them" (2006: 25). Thus, the killing of a snake could be seen as symbolising a wresting of control from the snake – not just control of him, but control of the environment in which he was so perfectly at home. Before the killing, the wilderness belonged to the snake, and afterwards it "belonged" to the human who killed him.

Eating snakes

Although AT hikers who killed animals on the trail cannot be described as having hunted them, it is apparent that through the deaths of some of these animals, the animal, as Marvin might put it, came to "belong" to the person responsible for their death. One way of explicitly demonstrating this fact was to eat the body of the dead animal. The narratives in which rattlesnakes were killed both mentioned that they were then eaten. Rattlesnakes are the most venomous, and consequently most dangerous, snakes on the AT. They are charismatic animals, known for their large size and distinctive rattling tails, and there is therefore an undeniable symbolism in the consumption of their flesh, linked directly to their perceived power and dangerousness. Levi-Strauss (1969: 162) famously wrote that animals make good totems "not because they are 'good to eat' but because they are 'good to think'". It appears that for certain hikers, snakes were good to eat *because* they were good to think.

The meat of animals is inherently symbolic even without being the flesh of a dangerous predator (see Adams, 1990; Fiddes, 1991). Fiddes (1991) describes meat's most important feature as the fact that it tangibly represents human control over the natural world:

> Meat satisfies our bodies but it also feeds our minds. We eat not only the animal's flesh; with it we drain their lifeblood and so seize their strength. And it is not only that animal which we so utterly subjugate; consuming its flesh is a statement that we are unquestioned masters of the world.
>
> (Fiddes, 1991: 68)

The killing and consumption of a rattlesnake can be interpreted as representing human control over even the most dangerous representatives of the natural world, and given that hikers are in a wilderness setting, can be viewed as representing control over the wild itself. For hikers who killed and ate rattlesnakes, their actions can be viewed as a symbolic 'conquering' of the wilderness that they are travelling through. Fiddes (1991: 33) writes that "an item's edibility depends not upon its flavour but upon its being found a position in our own classification of acceptable foods...we eat nothing in isolation, but as part of our culture – non-conformist habits and changes over time notwithstanding". Eating rattlesnake can be viewed as a demonstration of non-conformism by the hiker eating the snake, perhaps for the benefit of their own self-image, but undoubtedly also for the benefit of other hikers. Edensor (2001) has discussed considering tourism as a form of performance, which includes the opportunity for rebellious performance. His citation of Frykman and Lofgren's declaration that "regulation calls incessantly for freedom" (Frykman and Lofgren, 1996: 12, in Edensor, 2001: 61–62) is interesting because it brings up the question of how an AT thru-hiker carves a niche for themselves as a rebellious individual amongst other AT hikers. Many people who hike the trail like to think of themselves as rebellious, or at least, individualistic. This is reflected in statements about friends and family having been shocked at a blogger's plan to hike the AT, comments like "part of my reasons for hiking were rooted in a...contrarian nature", "one feeling I never get in the middle of the city, or in a town...I don't feel free", and "this morning...we took another step away from civilization". Hikers who want to stand out from the thru-hiker crowd as embodying individualistic traits might 'go the extra mile', so to speak, for example by killing and eating a rattlesnake. In Veeck's (2010: 249) research on people's consumption of "extreme" foods (such as reptile meat, amphibian meat or unusual body parts) she found that people who self-classified as "adventurous" tended to be attracted to unusual foods. Those of her study respondents who liked the idea of trying extreme foods associated them with "excitement, diversity and sophistication", and considered themselves as different from "the masses" (2010: 252). She writes that "some respondents seem to take perverse, machismo, and even defiant pleasure in transgressing the borderline between 'edible' and 'nonedible'" (2010: 252). Non-conformism is a highly valued trait among the AT thru-hiking community, and so the killing and eating of a dangerous predator not only demonstrates that the eater is at home in the wild – a "survivalist" – but also tangibly demonstrates that the eater is the ultimate non-conformist, as even most other hikers would likely shun, or be disgusted by, rattlesnake meat. Veeck (2010: 252) explains that neophilia – the love of novelty – is both an individual and a social trait. This type of action, then, is likely calculated to endow the hiker with a great deal of cultural capital; however, only among a certain sub-group of 'hard-core' thru-hikers. Those who, like the Trek editors, are adherents of the Leave No Trace philosophy promoted by the associations responsible for running and promoting the trail would find the act shockingly irresponsible and possibly even reprehensible (a case of moral disgust rather than physical disgust). Again, the hiker who kills and eats the snake has demonstrated their non-conformist status by shunning the 'rules' associated with hiking the AT. The reasons for this deliberate rule-breaking behaviour might be located in an individual's supposed need for self-actualisation, as described by Lyng:

> Simply put, people feel self-actualized when they experience a sense of direct personal authorship in their actions, when their behaviour is not coerced by the normative or structural constraints of their social environment. No longer impelled by intangible social forces, their actions reflect the immediate desires and goals of the ego.
>
> (Lyng, 1990: 878)

Thus, it might even be possible to consider that people who killed and ate rattlesnakes on the trail did so not only because they didn't care about Leave No Trace guidelines, but precisely because they felt an increased sense of personal autonomy by rebelling against them. Again, this type of deliberate rule breaking might also be seen as an appropriation of the environment of the AT, in opposition to the official 'managers' of the wilderness.

Although two examples of rattlesnake eating are not enough to form any reliable conclusions, it is interesting to note that the killers of the rattlesnakes were male, and at least in one instance, the person who prepared the flesh for eating was female. The blogger who rejected the snake flesh and chose not to eat it was also female. So, although those involved may like to view themselves as non-conformists, their performance around the killing, preparation, and eating (or rejection of) the rattlesnake can be seen as conforming to traditional gender stereotypes. If the "macho steak", as Fiddes calls it, denotes sexual prowess, then the body of a rattlesnake, in both shape and symbolism, represents a kind of sexual prowess on steroids.

The hiker "Golden" apparently contracted giardia soon after the rattlesnake killing, and had to stop hiking due to becoming very sick. As mentioned previously, giardia is contracted from water contaminated with the faeces of affected animals, and causes vomiting, diarrhoea, and severe stomach pains, amongst other things. A small minority of hikers – those who like to consider themselves in tune with 'the wild' to the degree that they consider their stomachs to be strong or immune to the potential illnesses in the wild – choose not to filter or otherwise decontaminate the water that they obtain from springs or ponds. Golden, described as a "natural survivalist" may possibly have fallen foul of the attitude of being so at home in the wild that he didn't need to filter his water.

In the only other narrative that described the eating of animals on the trail, a hiker noticed a line of ants trouping through his breakfast porridge, which after a brief moment of pause he continued eating, ants and all, having decided that they would make good extra protein. This scenario, very different in context from the killing and consumption of rattlesnakes, comes close to the sense of 'becoming wild' themselves that many hikers experienced while out on the trail.

Funny killing

The final blogs reviewed that mentioned the killing of animals included a hiker who sat on a millipede, accidentally killing the insect, and getting badly stung in the process, someone who found a deer tick attached to his knee, which he removed and "burned to a crisp", and a hiker who killed a rat in a shelter.

> So, I'm making a bit of a reputation for myself on the trail. After a 20+ mile day in the Shenandoahs, I arrived at the Blackrock Hut to see my friend Verruca there. As we set about eating our dinners, I see a long brown rat approach the shelter, observe us and then scurry underneath. It was about half an hour before we'd go to sleep. Even though I'm a deep sleeper, it's hard to get to sleep when you hear something skeetering about. As I turn off my light and the others sound to be asleep, I hear the pitter patter of rat feet. Spasmodically, I flip around with my light upon the rat, calmly eyeing me in the dark. Not tonight.
>
> I call down to Trekking Bro to fetch me my hiking pole. With some unsuccessful jabs, the rat sprints around the loft, but staying on the loft,

oddly enough. Hunching underneath the roof, I waddle to the kitty corner
section and find the rat. He observes me again and I him. Hovering my
pole four feet above his head I slam down hard. The rat crumples into
a fetal heap with permanent brain damage. Before any more suffering
I push a large stone to the head putting it to sleep.

The rat here is described in very similar terms to the mice that hikers
experienced feelings of disgust towards. There is a focus on how he moves;
he scurries, skeeters, pitter patters, and sprints. Like shelter mice, this rat has
invaded a domain supposedly reserved for human hikers, and this makes him
"matter out of place" (see Douglas, 2002 [1966]). Because the rat is out of
place in the human shelter, he is even more likely to be seen as dirty than if
he were in the woods. In Douglas's definition of matter out of place, dirt is
the by-product of our system of ordering and classification, and therefore a
relative idea: "shoes are not dirty in themselves, but it is dirty to place them
on the dining table; food is not dirty in itself, but it is dirty to leave cooking
utensils in the bedroom, or food bespattered on clothing..." (2002 [1966]: 45).
Yet rats already have dirty reputations, having carried the fleas that spread
the bubonic plague in the 14th century, and Edelman (2005: 124) points out
that rats have long been associated with sewer systems, and since the 19th
century, at times when the sewers would overflow their filthy contents into the
streets, the occurrence would be "paralleled in the imaginary world of horrors
to include a fear of a transgression of the separation of the upperworld and
the underworld". Kolnai (2004 [1929]: 57) describes rats as eliciting feelings
of anxiety, even of the "uncanny", partly due to their "dull, insidious character" and Jerolmack (2008: 86) points out that although rats no longer pose
the public health threat that they once did, "they have become culturally
enshrined as one of the most loathed animals on the planet". The double
whammy of being matter out of place *and* being viewed as a pestilent animal
makes the rat an example of what Knight (2000: 14) calls "*dirty* animals *par
excellence*", and as Jerolmack (2008: 86) describes, "beyond the sympathy of
most people". This status is arguably what contributes to him being seen as
killable. Yet this is not an emotionless evaluation, despite the nonchalance
that the blogger attempts to convey about his encounter. Marvin (2006: 25)
describes the killing of "vermin" as "hot deaths", in contrast to the killing
of domestic animals – "cold deaths", and the hunting of wild animals –
"passionate deaths". For Marvin, the killing of domesticated animals is generally premeditated, detached, and dispassionate, but the strong emotions of
irritation or disgust evoked by vermin/pests and their ostensibly transgressive
behaviour means that "it is the actual death, in and of itself, of the animal
that is wished for or desired, and there is relief and satisfaction when it is
accomplished" (2006: 18). Because rats are seen as polluting "when they are
simply present in places where humans think they ought not to be" (2006: 17),
killing the rat can also be seen as an act of cleansing. This path towards
being seen as killable is very different to the one by which snakes, especially

rattlesnakes, become killable – one situation motivated by an attraction to the animal (perhaps even a totemic attempt at assimilation) and the other a repulsion.

Klein (2015: 160) talks about how one of her AT informants knew another hiker named "Mouse Trap" who hiked with mouse traps and "cleared out" shelters with them. Ex-AT thru-hiker Bill Irwin wrote about the killing of shelter mice in a book about his hike. Most of the hikers he met had adopted what he called a "live and let live attitude" towards the mice, but others "were determined to leave the shelter with fewer rodents than when they arrived" (1996: 87).

> Ray and Louise travelled under the name Special Forces, and waged serious war against the mice. They carried traps and set them in strategic places almost immediately after arriving at a shelter. Before departing they would add several marks to the mouse body count in the register. One evening…there was a loud *snap* overhead, and a mouse caught in a trap fell right into Moleskin's macaroni and cheese! Louise grabbed the trap, held it up like a trophy and cackled, "Look at his little eyes!"
>
> (Irwin, 1996: 88)

As Irwin recounts more stories about mouse killing it is clear that they are meant to be taken as humorous.

> A father-and-son hiking team once found a mouse suspended in a shelter firepit with a hangman's noose and a neatly tied satin bow around his neck. It was the work of the Meyer's brothers, who had decided that "even a mouse deserves to be buried in a suit"… Every day or so they carved a tiny, wooden club about an inch and a half long with a hole in the end of the handle and a little string loop attached. After staying at a shelter, they left the little club hanging by a nail.
>
> (Irwin, 1996: 88)

A "set of instructions" would be left with the club, to strike a blow to the mouse's left temple, carry the mouse to a railway and lay him on the tracks, and then you would have "mouse jerky".

Jerolmack (2008: 79) cites Michael (2004: 258), who points out that humour is frequently used by people in making light of the death of "resolutely uncharismatic…clueless and stupid" nuisance animals, and states that it is, ostensibly, the low status of these animals that enables killing them to be funny. The "accidental" killing of the snake who slipped off of a hiking pole and into the fire became "a joke" when the hikers realised that they had killed a "harmless worm snake". Had the hikers killed a copperhead snake or a rattlesnake instead, the story would have taken on a different context and tone, in line with those species' perceived dangerousness, and therefore, their higher status. Also low in status (among the very lowest) are rats, and the hiker who blogged about his killing of one in the shelter he was staying in adopted a light-hearted tone to go with what he clearly saw as an uncontroversial kill.

Burt (Animal Studies Group, 2006: 198) argues in the conclusion to the Animal Studies Group's *Killing Animals* that "one kills in order to represent". Although not all the animal killings described in the narratives looked at in this chapter were carried out by the author of the narratives, all of the killers had witnesses to their actions – none of the reported killings were carried out when the hiker was alone. Some of these people then also blogged about their actions, meaning they could represent them to an even wider audience than their fellow hikers at the time. The killings are used to represent those having carried them out (and to an extent, their hiking partners, by association) in certain ways, and as part of the identity work of both long-distance hiking and blogging. What Burt refers to as "the fetishization of the documentation of killing" (2006: 198) points to the effectiveness of a story told about killing another being. In the handful of narratives about hikers killing animals on the AT, this effectiveness is accentuated by the unemotional way in which most killings were described.

Transgression

As with many uncomfortable encounters between humans and nonhuman animals that result in the animal being viewed as having transgressed in some way (see Turn, 2012, 2015; Jerolmack, 2008; Peace, 2002), animals on the AT were frequently written about as infringing upon human spaces, whether that be a mouse who chewed through canvas to enter a camper's tent, a rat who spent the night in a hiker shelter, or a snake whose 'transgression' was going to sleep in the warmth of a nearby firepit. Arluke and Sanders note the consequences to animals of being viewed as transgressors:

> Some animals...have a problem with their place in society. They may be freaks that confuse their place, vermin that stray from their place, or demons that reject their place. They are oddities that cause repulsion, unwelcome visitors that provoke fear, or dangerous attackers that arouse horror. In turn, society may ignore, marginalize, segregate, or destroy them.
>
> (Arluke and Sanders, 1996: 175)

The nonhuman animals talked about in this chapter transgressed in part because they did not meet hiker expectations about what animals on the AT should do or be like. Through their actions, and sometimes simply through their existence, they did not support the idea of being on a wilderness expedition (the way that charismatic animals like bears, moose, and eagles did). They did not reinforce the attitude of being on a pilgrimage (they were not 'welcoming' and 'beneficent' like deer, chipmunks, butterflies), and they were not 'cute' (like ponies and rabbits). What many of them did do, however, was to approach hikers, to take their food and destroy their belongings, to touch their skin, climb into their beards, sometimes to bite them, or try to latch onto them, to defecate on them, to drop onto their heads or shoulders from a tree branch, or sometimes simply to lie in the centre of the trail, directly in the path that the hiker wanted to take. Unlike cute animals, they did not appeal with their eyes or with gestures for the hiker to voluntarily give something; they just took it. Unlike wild animals that dwell in urban spaces, they did not flee from approaching hikers. It could be argued that the animals who hikers experienced uncomfortable encounters with were the species that demonstrated *the most* agency – ironically, the most self-willed-ness – the most wildness, if only in their behaviour towards hikers, which varied between avoiding them, ignoring them completely, or treating them as a resource. It cannot be coincidental that this type of behaviour coincided with feelings of fear and disgust on the part of the hikers who experienced it, hikers who were used to dwelling in a society in which humans have control over almost all interactions with other species, for which a lack of control will have been disturbing and even frightening.

Those species who treated hikers as a resource – most notably mice, but also ticks, leeches, and others – were seen as violating a hiker's boundaries. It is this violation of boundaries that most epitomises the way in which these animals were seen as transgressors. Douglas (2002 [1966]: 47) cites an essay by Sartre on viscosity, which Sartre described as repellent, saying that "there is no gliding on its surface. Its stickiness is a trap, it clings like a leech; *it attacks the boundary between myself and it*". Durham (2011: 138, my emphasis) says that "disgusting objects and acts are often those that violate boundaries, especially boundaries between what we consider human and non-human, sacred and profane, and between our selves and non-selves". She describes disgust as a feeling that emerges when a person imagines themselves as being "intimate" with something else. It should perhaps be added that this intimacy, in the case of hikers with disturbing animals, was involuntary and forced, which was essentially what made it disgusting.

Moral panic

Discussing the types of characteristics that make animals anomalous, Knight (2000: 15) cites "routine territorial boundary-crossing behaviour", arguing that "where such mobility occurs, the spatially based taxonomic order is threatened, with the result that a particular moral significance attaches to the animal in question". Jerolmack (2008) writes about pigeons living in cities who are considered to be transgressive, their primary offence being that they 'pollute' habitats which have been dedicated to human use. Like shelter mice, pigeons are considered to be matter out of place; animals who defy the human-drawn boundaries between spaces for humans and spaces for nonhumans. Jerolmack describes the reaction that ensues as "moral panic" (2008: 73), in reference to the sociological concept that looks at group social distress over perceived societal threats. In their writing on moral panics, Goode and Ben-Yehuda (1994) describe how certain phenomena (a person, group or even concept) can incite intense worry in members of a society, often when a rational assessment of the evidence points to the supposed threat being either non-existent, or far less harmful than the reaction to it would suggest. When the agents of perceived threat are persons or beings, they are known as "folk devils", are "stereotyped and classified and deviants", and can evoke "explosions of fear and concern" (1994: 149–150). Taking a constructionist perspective, Goode and Ben-Yehuda argue that, like all sociological phenomena, the threats that lead to moral panics are culturally constructed.

There is a long history of the moral panic that ensues from animal boundary-crossing resulting in the animals in question being deemed killable, for example, brown rats (*Rattus norvegicus*) emerging from sewers in England since the 1800s (Edelman, 2002, 2005), Chacma baboons (*Papio ursinus*) in South Africa's Cape, some of whom invade homes and take food from humans (Hurn, 2012), the urban pigeons (*Columbidae*) that Jerolmack writes about (2008), and even animals who have previously been awarded 'protected' status, such as the dingoes (*Canis lupus dingo*) of Fraser Island in Australia, who have been involved in altercations with humans at campsites (Peace, 2002). The actions of black bears on the AT, who also frequently 'invade' campsites – as described in Chapter 2 – can likewise be seen as having led to moral panic, as evidenced by hikers blogging about the numerous signs posted all over the AT about bear activity ("If you stopped for more than a few minutes a bear would come by and disembowel you. That was the vibe I got from the signs, anyway") and the seemingly overzealous activities of the National Park Service, and their program of trapping and removal or euthanasia of 'problem bears' and bears that have been mistakenly identified as problem bears (see WNCN, 2016). The attitudes of hikers to the animals that have been talked about in this chapter, however, cannot strictly be classified as fulfilling the theoretical model of moral panic, as there has been no group conception of a 'crisis' per se, and no institutionalised response to incidences of boundary crossing and other ostensibly transgressive behaviour (moral panic is typically characterised by what Goode and Ben-Yehuda term an "organized, movement-like activity" (1994: 167)). However, it is clear from hiker narratives about transgressive animals on the trail that many individuals experienced anxiety about the persistent boundary crossings of certain species, and that feelings of ambivalence and unease about these animals crossed over into individual experiences of a kind of moral panic.

Transgression all around

In a way, shelter mice are matter out of place wherever they are: inside shelters they are 'wild' animals who have invaded the domestic domain of hikers, yet outside of shelters they are not wild enough to fit in with the other true-wild animals living on and around the trail. They are doubly disgusting because of their ambiguity – they appear to fit nowhere. Jerolmack (2008: 75) lists some of the characteristics of so-called "problem" animals; they are not useful to humans (one might argue that all wild animals are not "useful" but on the trail animals like bears, moose, deer, and birds have symbolic uses for hikers), they are scavengers, they are uncharismatic or unattractive, they are perceived as damaging to human property, they prey on humans themselves, they are perceived as likely to spread disease. Although none of the animal species talked about in this chapter possessed all of these criteria, all of them possessed some.

If violating another's boundaries is a form of transgression then it can be argued that many humans on the trail also engaged in transgressive acts. The most extreme example of this would be the deliberate killing of animals, but picking up animals, and even approaching animals to take their photograph, also involves crossing the animal's boundaries. This happened not only with snakes, but frequently with animals viewed as cute and adorable. It may well be that while the hiker wrote about an engaging or appealing encounter, the nonhuman animal involved was experiencing their own uncomfortable encounter with the other. There is no reason for transgression only to be seen through the eyes of a human beholder, and lots of possibility for humans to have been viewed as 'pests' by the nonhumans they met.

Conclusion

This chapter has looked at the pestilence discourse around nonhuman animals on the AT. Up until this chapter we have explored how hikers who met other animals on the trail developed a deeper knowledge of the animals as beings. Yet this is not the case with encounters that were experienced as disgusting, frightening, or otherwise unpleasant. In these types of interaction, hikers tended to be too occupied by their own strongly negative reaction to the animal to be able to engage with the animal as an individual or as a species. Yet, as with other animal encounters, they were frequently interpreted as meaningful, or used as representative of something in the hiker's journey narrative. For example, they represented the difficulty and hardship that the hiker had to overcome on their journey, and the obstacles they had to face to obtain their goal. This type of 'meaning' may not have been conceptualised during unpleasant encounters, but was certainly present in the narratives subsequently written about them.

The introduction to this chapter raised a few questions that I wanted the chapter to address. Firstly, what it was like for hikers to live in close proximity to the kinds of animals that people usually avoid? It was often uncomfortable, and sometimes even disturbing, disgusting, and frightening. Yet, as with many aspects of life on the trail, people became accustomed to their close proximity to these animals, to being bitten, crawled on, and generally invaded. As their exposure to certain animals increased, so did their ability to tolerate the unceasing crossing of their boundaries by various species.

Secondly, what it was about these types of animals that led them to being perceived as distasteful, repellent or frightening? Many researchers argue that humans are predisposed to fear certain animals (snakes, rats) or feel disgusted in the presence of particular animal characteristics (sliminess, wriggliness) in order to protect ourselves from harm, either immediate or slow-burning. I have proposed that the increased vulnerability of long-distance hikers to being hurt or infected may have made them even more sensitive to feelings of disgust and fear elicited by particular animals.

Thirdly, how did hikers respond to these animals, and what were the consequences to the animal? The consequences for the animal of being perceived as disgusting, frightening, or both, were varied. Korsmeyer and Smith (2004: 1) write, "so strong is the revulsion of disgust that the emotion itself can appear to justify moral condemnation of its object – inasmuch as the tendency of an object to arouse disgust may seem adequate grounds to revile it". Likewise, according to Sherman and Haidt (2011: 3), "by inhibiting typical mentalizing processes, disgust may lead to the acceptance of treatment that would normally only be permitted in the case of objects or nonsentient beings". It may be that the idea of the disgusting/frightening animal as nonsentient – or less-than-sentient – is responsible for the fact that animals were hit with hiking poles, sticks, and stones, punched in the face, dropped in a fire, picked up and thrown, rushed at, yelled at, and killed. Yet people also frequently ran away from animals that they found disturbing, and in many ways 'cute' animals had their boundaries violated more regularly.

Finally, how did hikers make sense of these animal encounters in their blogs? Despite writing about trying to avoid interactions with certain animals, running away or otherwise ending interactions as quickly as possible, encounters with the disgusting and frightening were frequently written about in careful detail, and at length. There was a focus on the somatic aspects of the experience, both in terms of the bodies of animals (how they looked, how they moved) and in terms of the hiker's physical reaction to the animal (feeling chills, fear "pulsing" through their body, hairs standing on end, heart racing, stomach churning, and feelings of nausea). The detail and description used in hiker narratives seemed to indicate that although hikers disliked the encounter in the moment, they enjoyed – perhaps even relished – thinking about it afterwards. Hiker-bloggers seemed fascinated both by the animals in question, and by their own response to the animal.

References

Adams, C. J. 1990. *The Sexual Politics of Meat.* Cambridge: Polity Press.

Adkins, L. M. 2000. *The Appalachian Trail: A Visitor's Companion.* Birmingham: Menasha Ridge Press.

Angyal, A. 1941. Disgust and Related Aversions. *Journal of Abnormal and Social Psychology* 36: 393–412.

Animal Studies Group. 2006. Conclusion: A Conversation. In: Animal Studies Group (ed.) *Killing Animals*, pp. 188–210. Urbana and Chicago: University of Illinois Press.

Arluke, A. and Sanders, C. R. 1996. *Regarding Animals.* Philadelphia: Temple University Press.

Beeman, W. O. 1999. Humor. *Journal of Linguistic Anthropology* 9(1/2): 103–106.

Candea, M. and da Col, G. 2012. Introduction: The Return to Hospitality. *Journal of the Royal Anthropological Institute* 18(S1): S1-S19.

Curtis, V. 2013. *Don't Look, Don't Touch: The Science Behind Revulsion.* Oxford: Oxford University Press.

Davey, G. C. L. 1991. Characteristics of Individuals with Fear of Spiders. *Anxiety Research* 4(4): 299–314.

Davis, Z. 2012. *Appalachian Trials.* UK: Good Badger Publishing.

DeLoache, J. S. and LoBue, V. 2008. *Human infants associate snakes and fear.* Unpublished manuscript, University of Virginia, Charlottesville.

van Dooren, T. 2016. The Unwelcome Crows. *Angelaki* 21(2): 193–212.

Douglas, M. 2002 [1966]. *Purity and Danger.* London: Routledge.

Durham, D. 2011. Disgust and the Anthropological Imagination. *Ethnos* 76(2): 131–156.

Edelman, B. 2002. "Rats are people, too!": Rat–Human Relations Re-Rated. *Anthropology Today* 18(3): 3–7.

Edelman, B. 2005. From Trap to Lap: The Changing Sociogenic Identity of the Rat. In: J. Knight (ed.) *Animals in Person: Cultural Perspectives on Human–Animal Intimacies*, pp. 119–139. Oxford: Berg.

Edensor, T. 2001. Performing Tourism, Staging Tourism: (Re)producing Tourist Space and Practice. *Tourist Studies* 1(1): 59–81.

Fiddes, N. 1991. *Meat: A Natural Symbol.* London: Routledge.

Frykman, J. and Lofgren, O. 1996. Introduction. In: J. Frykman and O. Lofgren (eds.) *Forces of Habit: Exploring Everyday Culture.* Lund: Lund University Press.

Goldenberg, J. L., Pyszczynski, T., Greenberg, J. and Solomon, S. 2000. Fleeing the Body: A Terror Management Perspective on the Problem of Human Corporeality. *Personality and Social Psychology Review* 4(3): 200–218.

Goode, E. and Ben-Yehuda, N. 1994. Moral Panics: Culture, Politics, and Social Construction. *Annual Review of Sociology* 20: 149–171.

Herz, R. 2012. *That's Disgusting.* New York: W. W. Norton & Company, Inc.

Hurn, S. 2012. *Humans and Other Animals.* London: Pluto Press.

Hurn, S. 2015. Baboon Cosmopolitanism. In: K. Nagai, K. Jones, D. Landry, M. Mattfeld, C. Rooney and C. Sleigh (eds.) *Cosmopolitan Animals*, pp. 152–166. Basingstoke: Palgrave Macmillan.

Ingold, T. 2013. Prospect. In: T. Ingold and G. Palsson (eds.) *Biosocial Becomings: Integrating Social and Biological Anthropology*, pp. 1–21. Cambridge: Cambridge University Press.

Irwin, B. 1996. *Blind Courage.* United States: WRS Publishing.

Isbell, L. 2006. Snakes as Agents of Evolutionary Change in Primate Brains. *Journal of Human Evolution* 51: 1–35.

Jerolmack, C. 2008. How Pigeons Became Rats: The Cultural-Spatial Logic of Problem Animals. *Social Problems* 55(1): 72–94.

Klein, V. A. 2015. The nature of nature: space, place and identity on the Appalachian Trail. Unpublished PhD thesis, Kent State University College, Ohio, US.

Knight, J. 2000. Introduction. In: J. Knight (ed.) *Natural Enemies*, pp. 1–35. New York: Routledge.

Kolnai, A. 2004 [1929]. *On Disgust*. Illinois: Carus Publishing Company.

Korsmeyer, C. and Smith, B. 2004. Visceral Values: Aurel Kolnai on Disgust. In: A. Kolnai, *On Disgust*, pp. 1–22. Illinois: Carus Publishing Company.

Lee, S. W. S. and Ellsworth, P. C. 2013. Maggots and Morals: Physical Disgust Is to Fear as Moral Disgust is to Anger. In: K. R. Scherer and J. R. J. Fontaine (eds.) *Components of Emotional Meaning: A Sourcebook*, pp. 271–280. Oxford: Oxford University Press.

Levi-Strauss, C. 1969. *Totemism*. Bucks: Pelican Books.

LoBue, V. and DeLoache, J. S. 2008. Detecting the Snake in the Grass: Attention to Fear-Relevant Stimuli by Adults and Young Children. *Psychological Science* 19(3): 284–289.

Logue, V. and Logue, F. 2004. *The Appalachian Trail Hiker*. Birmingham: Menasha Ridge Press.

Lorimer, J. 2017a. Probiotic Environmentalism: Rewilding with Wolves and Worms. *Theory, Culture and Society* 34(4): 27–48.

Lorimer, J. 2017b. Parasites, Ghosts and Mutualists: A Relational Geography of Microbes for Global Health. *Transactions of the Institute of British Geographers* 42(4): 544–558.

Lyng, S. 1990. Edgework: A Social Psychological Analysis of Voluntary Risk Taking. *The American Journal of Sociology* 95(4): 851–886.

Marvin, G. 2001. The Problem of Foxes: Legitimate and Illegitimate Killing in the English Countryside. In: J. Knight (ed.) *Natural Enemies: People–Wildlife Conflicts in Anthropological Perspective*, pp. 189–211. London: Routledge.

Marvin, G. 2006. Wild Killing: Contesting the Animal in Hunting. In: The Animal Studies Group (eds.) *Killing Animals*, pp. 10–29. Chicago: University of Illinois Press.

Matchett, G. and Davey, G. C. L. 1991. A Test of a Disease-Avoidance Model of Animal Phobias. *Behaviour Research and Therapy* 29(1): 91–94.

Merleau-Ponty, M. 2014 [1945]. *Phenomenology of Perception*. Oxon: Routledge.

Michael, M. 2004. Roadkill: Between Humans, Nonhuman Animals, and Technologies. *Society & Animals* 12(4): 277–298.

Mulkens, S. A. N., de Jong, P. J. and Merckelbach, H. 1996. Disgust and Spider Phobia. *Journal of Abnormal Psychology* 105(3): 464–468.

Mundkur, B. 1994. Human Animality, the Mental Imagery of Fear, and Religiosity. In: T. Ingold (ed.) *What Is an Animal?*, pp. 141–184. London: Routledge.

Öhman, A. and Mineka, S. 2003. The Malicious Serpent: Snakes as a Prototypical Stimulus for an Evolved Module of Fear. *Current Directions in Psychological Science* 12(1): 5–9.

Peace, A. 2002. "The Cull of the Wild". *Anthropology Today* 18(5): 14–19.

Rozin, P. and Fallon, A. E. 1987. A Perspective on Disgust. *Psychological Review* 94: 23–41.

Sherman, G. D. and Haidt, J. 2011. Cuteness and Disgust: The Humanizing and Dehumanizing Effects of Emotion. *Emotion Review* 3(3): 1–7.

Song, S. H. 2001. The Great Pigeon Massacre in a Deindustrializing American region. In: J. Knight (ed.) *Natural Enemies: People–Wildlife Conflicts in Anthropological Perspective*, pp. 212–228. London: Routledge.

Strohminger, N. 2014. Disgust Talked About. *Philosophy Compass* 9(7): 478–493.

Veeck, A. 2010. Encounters with Extreme Foods: Neophilic/Neophobic Tendencies and Novel Foods. *Journal of Food Products Marketing* 16(2): 246–260.

Wilson, E. O. 1984. *Biophilia*. Cambridge: Harvard University Press.

WNCN. 2016. *Wrong Bear Was Killed after Attack on NC Hiker, Officials Say*. Available at: http://wncn.com/2016/05/23/wrong-bear-was-killed-after-attack-on-nc-hiker-officials-say/

5 Chirps, quacks, croaks, howls, and "what was that?"

Animal sounds on the trail

I go up over logs slowly
On my painfully reborn legs
My ears putting out vast hearing
Among the invisible animals

(Dickey, 1996: 86)

The AT is frequently referred to by hikers as "the Green Tunnel", due to much of it being enclosed by dense tree cover, which makes it difficult for people to see far ahead or around themselves. As one blogger wrote: "we are constantly in the 'green tunnel' and the views are few and far between". This means that hikers frequently encountered other animals through hearing them, rather than seeing them – Dickey's "invisible animals" (1996: 86) – or through hearing them before they saw them. In addition, many of the other-than-human animals living on and around the trail are more active during the night, again meaning that hikers could hear their comings-and-goings often without being able to see the animal who was making the sounds that they heard. Anthropologists studying people who inhabit dense forests have noted the significance of sound in their lives, in terms of knowledge and meaning making, bodily orientation within the environment, and relationship to other inhabitants (see, for example, Feld on the Kaluli people of Papua New Guinea, 1996; Schieffelin also on the Kaluli, 2005 [1976]; Gell on the Umeda, 1995). So, too, the auditory perception of animals on the trail was always vital to, and indivisible from, people's experiences of them.

> While hiking north from Thomas Knobb Shelter the ponies started getting closer, but the fog was so thick that for a couple of miles we could only hear them whinny to each other.

The composer R. Murray Schafer (1994: 7) coined the term "soundscape" to describe what he called an "acoustic environment" that could be isolated and studied "just as we can study the characteristics of a given landscape". Given that the AT is more than 2,000 miles long, it is important to note that there is no homogenous 'AT soundscape', and that the acoustic characteristics of the trail will vary according to where on the trail a hiker finds herself, as well as what season it is, what time of day, and so forth. In general, however, bloggers talked about the sounds of 'nature' being far more dominant on the trail than they were used to from their home communities, and about human-generated sounds being far less dominant than they were used to – and sometimes almost completely absent.

One person wrote a lengthy post about sounds on the AT, which will be useful to look at as it provides a detailed description of how sound could be experienced on the trail, as well as introducing some of the key themes arising from hiker narratives.

> [I]t seems only natural that we speak, write and remember the Appalachian Trail in predominantly visual language. For instance, it is a national scenic trail, and Benton MacKaye described the purpose of hiking it as "to walk, to see and to see what you see". Yet one of the overwhelming experiences of hiking on the Appalachian Trail is the aural or sonic qualities of the trail environment.
>
> Most people heading to the trail today live in an urban or suburban environment where extraneous artificial noise (sirens, car alarms, trains and airplanes) are a normal part of the outdoor soundscape. The Appalachian Trail's ancient, natural soundscape is encroached upon by urbanizing noise pollution in many places, but if you walk far enough you will find space and time to absorb the horror which is silence, or near silence. I say "horror" because most urbanized people getting into the back country react to this silence by talking (to oneself or strangers), hooting and howling (because all of us are primates), or strapping on headphones and listening to music or talk radio or audiobooks. The initial reaction to the sound of nature is to block it out. In cities we must learn not to listen or be overwhelmed. In the woods, it takes time to unplug...

In describing certain human-made sounds as "artificial noise" and "noise pollution" the blogger echoes the sentiments of many hikers who expressed disappointment at still being able to hear human-derived sounds like busy roads or airplanes, at certain points on the trail. There was certainly an expectation among hikers, directly linked to the idea that being on the trail was being in the 'wilderness', that while on the trail they should be able to hear only 'natural' sounds, not those generated by humans – what Krause (2013) refers to as *anthrophonic* sounds – and particularly not those generated by human-made machinery. Although human-generated sounds were far less common than at home, this sometimes meant than almost any human-made sound (or at least, any not coming organically from a human body) was considered an intrusion. The exception to this rule – at least for some people – were the sounds that could be listened to through headphones. As the blogger here puts it: music, talk radio, audiobooks. These are described, presumably tongue-in-cheek, as being used by some to block out the "horror" of the woodland silence. Yet the blogger corrects himself; "silence" becomes "near silence", which is then described as "the sound of nature". What the blogger apparently means by "silence", then, is an absence of human-derived sounds.

> The society addled hiker can hear, if he or she listens, the voices of thousands upon thousands of birds, insects – and larger animals. The creeks and stones speak too. The wind and the sky have a voice, and it is the loudest voice because it speaks with water…on a long hike, after the hiker is talked out, sick of all her music and too tired to dwell in the past, the sound of the woods finally penetrates and saturates the human psyche. I would suggest that there is an ancient (proto) human in each and every one of us that is pleased by the immersion of the psyche in the sounds of nature; and a long, immersing wilderness experience can have a profound effect on the noise inside a person's head. The sound of sanity is silence. The sounds of nature, to borrow a phrase from Robert Bly, are "news from the universe" … One natural sound of the trail I remember most often is the sound of the summer woods at night – so many nonhuman voices. I remember, too, the rather sobering sound of the Smoky Mountains at high altitude in a late winter snow storm: utter deathly silence interrupted by howling wind. I recall hearing the voices of hawks and owls; the call of Canada Geese in New Jersey glacial swamps, or loons in Maine; coyotes and crows. The sound of the rain is more pleasant than the mire of a rainy trail. The sound of water dominates the Appalachian Trail, whether it is in the air feeding storms, falling as rain or sleet or snow, the relentless sloshing of hiking boots in wet leaves and puddles – and waterfalls. To be absorbed in this aural environment is to be washed of the madness of the overpopulated, artificial urban hive… Along parts of the AT in Maine, the terrain is remote enough so that the sound of an airplane passing overhead is startling.

The blogger again equates an absence of anthrophonic sounds with "silence", yet his description of the sounds made by "thousands upon thousands" of animals, as well as the wind and rain, and his own feet sloshing through wet leaves and puddles conjures up a sonic environment rich in diversity, and composed primarily of what Krause (2013) terms *biophonic* sounds (the sounds of living organisms) and *geophonic* sounds (nonbiological natural sounds, like wind and rain). For the blogger, being present in this so-called soundscape is being "absorbed", an immersion that allows people to be "washed of the madness" of human society; a phrase very similar to some of the statements made by hikers who considered themselves to be on a pilgrimage. His association of a late winter snowstorm with the Smoky Mountains, of Canada Geese with New Jersey and loons with Maine speaks to the place-making effect of sound. To be, as he puts it, "absorbed in this aural environment" is to be taken up within a sonic landscape, even to become a part of it.

> On my 2011 hike from Georgia to Maine, I accidentally erased the contents of my MP3 player. I was hiking through Fahnestock State Park in New York, delirious from the heat and my sweaty, dirty fingers pressed buttons they should not have pressed. All my jams – gone in an instant. I almost cried, but I was sick of all my music anyway, and in a few hours I forgot about all that stuff. It was pleasant to have technology to lean on for the company of comforting sound, but if a hiker stays out long enough and acclimates to the natural aural environment, the mountains and forests will provide their own comforting soundtrack. I was lucky to camp with Fruitbat…in Georgia in 2011. An ecologist, he could hear bird and insect sounds in the woods, identify the species and provide a wealth of information about what we were hearing. Fruitbat was a guy who could listen to the woods and pick up many more 'channels' from Mother Nature's broadcast than most people I knew.

The narrative concludes with an anecdote about how the blogger was forced to go without his music – the "company of comforting sound" – and so discovered that the woods provided their own sonic comforts, in the form of bird and insect sounds. In DeLuca's (2016: 88) research on people listening to wolves in the Isle Royale National Park near the US–Canada border, the author describes the effect on people of the lack of familiar sounds, remarking that within this "silence", "park visitors experienced a noted switch from passive hearing to active listening". This type of listening seems to be what the blogger is describing, particularly when he talks about having camped with another hiker, Fruitbat, and being impressed with his ability to listen to the woods, and identify the animals who were making sounds. It is apparent from blogger narratives about animal sounds that knowing the name or 'identity' of the animal producing the sound was a different listening experience to not knowing what type of animal was producing it.

Scholars of sound and listening have identified different modes of listening, with many putting forward their own taxonomies of listening (see Rice, 2015 for some prominent examples). Rice (2015: 99) writes that "a person may listen to something intently, absorbed in the sound, but distracted, indifferent, deconcentrated or even unconscious listening are also possible". The intent, active listening, or "deliberate channelling of attention toward a sound" (Rice 2015: 99) that Fruitbat engaged in to identify species by their voices was a mode of listening recognisable across many hiker narratives.

National Park Service (NPS) Natural Sounds Program

On long sections of the AT, the sounds that hikers heard had, to at least some limited extent, been 'managed', or "acoustically designed" (see Schafer,

1994: 271). The US NPS, which is responsible for the management of much of the land that the AT traverses, runs the Natural Sounds Program aimed at conserving the unique sound environments of the national parks. According to the programme, the characteristic sounds of each park are viewed as natural resources.

> Each national park is unique – for the diversity of wildlife, beautiful landscapes, historic resources, and also variety of sounds. The natural and cultural sounds in parks awaken a sense of wonder that connects us to the qualities that define these special places. Park sounds are part of a web of resources that the National Park Service protects under the Organic Act.
>
> (NPS, 2017: np)

The programme website (nps.gov/subjects/sound) describes natural sounds as vital to ecosystems (for animal communication, territory establishment, mating, and predator avoidance) as well as to visitor experience of parks. The programme also incorporates "cultural sounds" as protected resources, including Native American music, train whistles, and mission bells. The 'historical' or 'traditional' status of these anthrophonic cultural sounds appears to exempt them from being thought of in the same way that music from someone's iPad, or the sound of aircraft flying overhead, would be. The protection of cultural sounds by the NPS is indicative of a change in attitude towards 'wilderness', which for much of recent history was only seen as authentic in the absence of humans and human artefacts.

The sounds of nature, uninterrupted by human-generated sounds, are described on the Natural Sounds Program website as "natural quiet" (see Dumyahn and Pijanowski, 2011; Pilcher et al., 2009), in a similar manner to how the hiker quoted above refers to the absence of anthrophonic sounds as "silence". Yet the notion of natural sounds as quiet or silent seems to involve a denial or lack of recognition of the complexity and richness of information transferred through nonhuman sounds, for those who know how to interpret them. The website also comments on how detrimental noise is to human visitors as well as to the nonhuman animals living in national parks (see Passchier-Vermeer and Passchier, 2000; Pilcher et al., 2009; Staples, 1996), and how work is carried out to minimise noise intrusions (the Quiet Parks Program, www.nps.gov/articles/keeping-the-peace-and-quiet.htm).

The website includes a 'sound gallery', which features recordings of a multitude of individual animal species, as well as meteorological (rain, thunder), geological (avalanche, rockfall), 'human-caused' (boat, chainsaw), and 'cultural/historical' (battle sounds, cannon fire) sections, but does not offer any 'soundscape' or environmental recordings. Indeed, the sounds in the gallery have all been abstracted from the environment in which they occurred, a common convention in sound and music recording (see Bruyninckx, 2011). Yet, as Fisher (1999) notes, when we hear the sounds of animals in real life, we tend to hear them as *part* of an ensemble of sounds in a given sonic environment. He argues that part of the pleasure of hearing natural sounds, including animal voices, is to hear them in this way, backed by the other sounds of the environment.

Chirps, squawks, songs

The most written-about animal sounds on the trail were those made by birds. Schafer (1994: 9) talks about what he terms the "keynote" sounds of a place as sounds which are ubiquitously there, setting the fundamental tone of the sonic environment, around which all other sounds can modulate. From hiker narratives it seems apparent that birds (along with insects, wind, and water) were responsible for the keynote sounds of the trail. People blogged that "the birdsong is so beautiful...especially the way in which the melodies weave in and out with the rustling of the leaves and flowing of the streams", and that "there is nothing like spending the night in nature and waking up in the beauty of a sunrise and the chirping of birds outside your door". Several bloggers talked about how much they liked going to sleep to the sound of owls, loons, and other bird species. Although Schafer describes keynote sounds as not always being consciously heard, the fact that the sound environment of the trail was so radically different to that of most hikers' everyday lives is likely to be the reason for these sounds being actively listened to, and written about.

Some people found humour in the perceived strangeness of certain sounds made by birds. One blogger wrote that she "saw three wild turkeys on this day, which is so delightful. Turkeys make sounds that are simply hilarious", and "we heard a wild turkey doing a terrible job of being discreet, but it was pretty fun trying to locate it". Others ascribed comedic personalities to the birds around them.

> Birds sing non-stop. Whip-poor-wills, wood thrushes, and finches are common, as are the barred owls who we call Party Owls because they always get rowdy right around the time you want to go to sleep. The Eastern Wood-Pewee is another favourite. When it calls it sounds like, "Hey guys...?! Hey guys...?! Awwwww..." (disappointed). Poor lil' Pewee. Makes me laugh every time.

Yet not all experiences of bird sounds were delightful. When a couple of hikers were confronted by a mother grouse protecting her young, one wrote that the sound she made caused "shivers coursing through my body from head to toe and back again. That hiss will live on in my nightmares for a while".

Attending to birds

Birdsong has been repeatedly identified as the sound, or one of the sounds, that humans most like to hear in their environment (see Dumyahn and Pijanowski,

2011; Schafer, 1994). For Schafer, "no sound in nature has attached itself so affectionately to the human imagination as bird vocalizations" (1994: 29). Carles et al. (1999) cite experiments carried out by Bjork showing that the sounds of birdsong are better than any human-generated sounds at inducing states of relaxation. Of course, the comical and frightening experiences of birds cited above are evidence that bird vocalisations do not automatically induce feelings of relaxation in human listeners. These birds could perhaps be seen as being noncompliant with expectations about 'soothing' vocalisations; a reminder of their autonomous Otherness.

During Whitehouse's study of individuals' relationships with birds through sound (2015), he received several stories from people talking about their own experiences of listening to birds. One of these people, a farmer, described listening to the dawn chorus every day as giving him a "real lift". For Whitehouse, "this lift...stems from a sense of resonance that comes from this sympathetic attention to the activities of other beings around us" (2015: 64). The idea of this kind of satisfaction in the sounds of other beings goes far deeper, then, than the enjoyment of a pleasant keynote sound, or feelings of tranquillity. Whitehouse is proposing that human lives are enriched when we acknowledge and attend to – when we witness – nonhumans going about their lives. This echoes Bernard Williams's (1992) view, cited by Fisher (1999: 34), that it is the "otherness" of nature that makes it so valuable to us. As Fisher points out elsewhere, birds are generally not significantly present in the visual landscape (1998: 168). Thus, for hikers on the AT, listening to birds was the primary way of getting to know them.

A little night music

> We went to sleep to the sound of crickets and cicadas. Much later in the evening, we enjoyed a lovely chorus of barred owls as they called back and forth to one another, "who cooks for you? Who cooks for you?" – which is what their call sounds like. I love the sound of the woods at night. It's like music to my ears.

Hiker narratives described the woods as being alive at night with the sounds of animals going about their business. They wrote about hearing mice scuttling around inside shelters or outside tents, "random intervals of croaking frogs", and ducks quacking "sleepily" on a nearby lake, dogs barking, caterpillar poo falling like rain on tents, bears padding around campsites, and coyotes howling in the distance. One hiker commented, "There's some noisy animals outside, and we all get tucked further into our sleeping bags after shining our headlamp out onto a possum". The sounds of insects often dominated.

> Every seventeen years the cicadas come out to mate and die. This brood is called Magicicada. This horrifying sound of screeching fills the night air. They sound like the main title track from The Predator. I'm from

Wisconsin and I've never heard this. Other layers of night noises are crickets, something that sounds like a faraway alarm clock, trains coming every 40 minutes, and my heart thumping.

This blogger's conflation of electronic and natural sounds is reminiscent of Smith's (2011: 48) description of how early American industrial workers "braided" the sounds of the countryside and the sounds of industry together, sometimes perceiving natural sounds as industrial ones, and vice versa, rather than perceiving a juxtaposition between the nature and machine sounds.

As discussed in Chapter 2, the sounds of bears moving around at night could feel very threatening to hikers.

So a bear decided to sniff me out last night, that was scary but I somehow managed to fall asleep afterwards. It sounded like a deep, throaty, growly sniff. I'm surprised it didn't scare me more than it did, but I refused to turn on my head lamp and see it. As if, by not seeing it, it was less real than it actually was.

At night, even more so than during the day, hearing animals going about their lives in an environment that belonged to them evoked in hikers a recognition of their Otherness, a recognition that could be reassuring, invigorating, and sometimes very frightening. The hiker who chose not to turn on a light and look at the bear sniffing around her tent thought that seeing the bear would make him too "real", demonstrating an interesting perspective on which of her senses she trusted the most ('seeing is believing', as some might say). Yet on many other occasions people expressed fear in large part because they were unable to see what animal was producing the sounds that they heard.

What was that?!

It is generally accepted that the perception of sound is likely affected by visual attributes (see Matsinos et al., 2008). On occasion, particularly during the night, hikers would hear animal sounds that they weren't able to identify, because they couldn't see the animal producing the sounds. Their narratives about these experiences clearly show their discomfort, and even sometimes fear, at not knowing what kind of animal they were hearing. Without any visual frame of reference or recognition, the animal's Otherness could be experienced as overwhelming. One person described waking in the night to a "rumbling" sound, which she realised was coming from a small animal caught between her tent and the rain-fly, but was overcome with panic when she could not immediately tell whether it was a bird, bat, or mouse. Another wrote about the fear she felt when woken by noises outside of her tent.

I know the silence of the night amplifies sounds. I know distance is easily misjudged out in the wilderness. I know tiny creatures can produce sounds

1000 times their size. I know imaginations run wild. I know it probably was just a late night chipmunk happily bounding through the woods, rushing home to make curfew. But to me, it sounded like a huge elephant with a machete foraging through the bush right next to our tent…

Some bloggers who chose to go night hiking described their experiences of "animal noises" as they made their way along the trail. One wrote about having recently read an article on mountain lions returning to New Hampshire, the state that she was hiking through, and being "spooked beyond belief" when she heard animal sounds.

I called out, in an attempt to scare them, only to hear some people respond in return. I had never been so relieved to hear people before… We both thought we were scary animals and were both relieved that we were not the dreaded mountain lion…I said farewell and continued on…I sang loudly the entire way to spook off any possible mountain lions.

The necessity of listening out for others rather than looking out for them is something that has been noted by anthropologists who have studied people who live in dense forest environments. Gell (1995: 235), writing on the Umeda people of New Guinea, argues that the forest environment "imposes a reorganization of sensibility", so that hearing becomes the prioritised sense, a transformation that "has manifested consequences in the domain of cognition". For the Umeda, an animal is classified and recognised through the sounds that they make – *not* through the ability to see what they look like. For thru-hikers, more attuned to the sight-centrism of 'Western' culture, not being able to observe the animal source of sounds was disconcerting.

Even during daylight animal sounds could sometimes be difficult to identify. Many bloggers mentioned hearing an insect-like sound that turned out to be a rattlesnake, for example one who heard "what sounded like a million cicadas coming alive", before realising the sound was coming from a rattlesnake in her path. For some, rather than being warned off by the rattle, they followed the sound out of curiosity to try to find the snake that was making it – the opposite of what the snake would have intended. Another talked about finally knowing the source of a mysterious sound.

The sound of a grouse taking flight catches my ear, and I feel enlightened that I can actually recognise the powerful rumbling sound of it taking flight. For at least the first two months of being on the trail, I would hear a deep humming starting slowly and then increasing its tempo rapidly until it faded away into the ambience of the woods. The source of this mysterious sound eluded Scone Boy and I for weeks on end. Several times a week we would hear it penetrating the quietness of the woods, and we'd stop suddenly to face each other with eyes widened, then we'd look around frantically in an attempt to calculate the direction of it. We pondered what it could be endlessly, as we always heard the grouse taking flight without actually seeing it. I even conjectured it could've been an earthquake at one point, which is rather embarrassing considering it's just an oversized chicken flapping its wings.

For this hiker, the experience of listening to a "powerful rumbling sound" without knowing its source was markedly different to the experience of listening to a "grouse taking flight". This type of recognition could perhaps be likened to Clarke's (2005) argument regarding listening to music, that a listener's prior musical knowledge is integral to their current listening experience, and therefore, as Rice points out, "the sensory dimension of listening... might be understood as only one aspect of its wider cognitive and affective engagements" (2015: 103). Before being able to identify the sound as a grouse, the blogger and his hiking partner clearly felt a sense of threat in the sound that they heard. Pickering and Rice (2017) describe ambiguous sounds – those that could have more than one interpretation – as demanding attention, and provoking a human need for identification and categorisation. They note that ambiguous sounds have an innate "propensity to be felt as dangerous" (2017: np). For hikers, part of the stress of ambiguous sounds was the inability to identify whether an animal was known or unknown, harmless or harmful.

Hiker sounds

Krause (2013) describes *anthrophony* – human-generated sound – as comprising four basic types: electromechanical sound (which, in terms of sounds that might be heard on or from the AT, could include aircraft, cars, quad bikes, tractors), physiological sound (breathing, talking, yelling at bears), controlled sound (music played through headphones, drumming), and incidental sound (footsteps that crunch on leaves or slosh through mud, a tent flap being zipped up, the rustle of a sleeping bag). His exclusion of human-generated sounds, even physiological ones such as breathing and talking, from the category *biophony* (which he describes as "the sounds of living organisms") is characteristic of the human/nonhuman animal divide in contemporary thinking. It also fits well with the historically prevalent view of wilderness as a place that humans visit but do not dwell in, which encourages people to think of human sounds (including their own) as intrusions into the 'natural soundscape'. Yet the sounds that hikers themselves made informed part of their emotional response to the environment. As Feld (1996: 97) writes: "sound, hearing, and voice mark a special bodily nexus for sensation and emotion because of their coordination of brain, nervous system, head, ear, chest, muscles, respiration and breathing". Hiker sounds also inevitably formed part of the so-called soundscape of the trail. Indeed, Abram describes human voices as part of an animate environment:

Language as a bodily phenomenon accrues to all expressive bodies, not just the human. Our own speaking, then, does not set us apart from the animate landscape, but – whether or not we are aware of it – inscribes us more fully in its chattering, whispering, soundful depths.

(Abram, 1994: 121)

Nevertheless, it was generally acknowledged among bloggers that the less sound a hiker made, the more likely they were to encounter other animals. In a post entitled "5 Easy Tips to See More Wildlife on Your Thru-Hike", the author recommends that hikers add rubber tips to their hiking poles to make their hiking quieter, and that they hike quietly alone rather than chatting with fellow hikers. Others, while disappointed at not having seen many animals, acknowledged that their "chatter and noisy steps" were likely to be deterring wildlife.

There were times when hikers deliberately used sounds to try to communicate with other animals on the trail. The official advice from the NPS on dealing with black bears was to make a lot of noise to try to scare them away, and several hikers wrote about yelling at bears, clicking their hiking poles together, blowing bear whistles and making noise with whatever they had to hand. Others tried to engage in more friendly communication with animals, sometimes with limited success.

> To my right I heard what sounded like a very agitated bird squawking at me...I whistled a soothing tune in its direction (think Disney) and imagined it whistling back to me and flying by my side singing game show theme songs all day. That is not what happened. Instead, it angrily chirped, flew right past my head, and landed on a branch not far in front of me. It decided to turn and fly directly at my face, only swerving at the last second...

The blogger's failure to communicate effectively with the agitated bird, and her seemingly aggressive response to his whistling, forced him to recognise the difference of the bird's perception of the encounter to his own, and the bird's inherent Otherness, which would not comply with his expectations from the encounter.

These bloggers' heightened awareness of their own sounds in the woods is reminiscent of Gell's (1995) description of the forest-dwelling Umeda, who have a very deliberate awareness of the sounds that they make as they move through the forest. Gell describes the Umeda as being highly disciplined about noise; either they would be very quiet, or they would be very noisy, but they would not tolerate "intrusive unsocial background noise" (1995: 238). The more time that hikers spent on the AT, the more attuned they too became to their own – and other hikers' – sounds.

Anthrophony and noise

The use of headphones (as well as other technological artefacts, like smart phones and iPads) is a debated topic among hikers on the trail, given that they can be considered as inappropriate to a 'wilderness' environment. As Rice notes: "listening practices...can serve as indicators of moral, social, civic, psychological, or even spiritual well-being or decline" (2015: 103). There was a sense among some 'purists' that listening through headphones instead of attending to the sounds of nature could be associated with this spiritual "decline".

In the previously mentioned post offering tips for seeing more wildlife, the blogger recommended putting headphones away.

> Unplug. While you may need your tunes, podcasts or books on tape to get you up some of the mountains, when you plug into technology you miss out on a lot of what's going on around you. Slight rustles and warning calls can give you information about where to look before an animal runs off and blends back into the forest background, and you'll miss those subtle signals if you have your headphones turned up.

Another hiker, who narrowly missed stepping on two rattlesnakes, advised that hikers "keep one earbud out!" so that they could hear the warning rattle. This kind of compromise solution draws attention to the fact that hikers actually need to be able to listen to their surroundings for their own safety – indeed, Rice (2015: 101) points out that listening is a valuable survival skill when considered from an evolutionary perspective – and yet also demonstrates that some are so dedicated to listening to whatever is coming through their headphones that they would rather half-listen to the headphones and half-listen to their surroundings than turn the technology off. This resultant split listening, in which the listener occupies two sonic worlds at the same time, or is split between them, has previously been observed as characteristic of how people move through contemporary urban environments, and has been termed "periscopic listening" (see Drever, 2014).

Bull (2000: 2), in his ethnography of people who habitually use "personal stereos" as they move through urban environments, describes users as making "attempts at creating manageable sites of habitation". He argues that the use of personal stereos in constructing "familiar soundscapes" helps users to move through urban environments while never leaving "home", and to maintain a sense of their own identity within an "often impersonal environment" (2000: 24). Although Bull's focus is on urban environments, the AT is another type of unfamiliar, and sometimes impersonal-feeling, environment. The trail can feel like a strange and bewildering place to hikers, and Bull's findings point to a reassuring sense of 'home' that familiar tracks on an iPod or smartphone can provide. Yet Bull also notes the different type of relationship that will be experienced with others when listening to music through headphones:

> Personal stereos tend to be non-interactive in the sense that users construct fantasies and maintain feelings of security precisely by not interacting with others or the environment. Users rather construct a

range of interpersonal strategies that are inherently asymmetrical. Ways of auditized looking are developed which are inherently non-reciprocal, functioning to bolster the user's sense of power and control in urban space. Users often approach the public one step removed and this affects how they interact and respond to situations that confront them.

(Bull, 2000: 25)

From Bull's observation, it is clear that while the use of headphones can offer a sense of control and home-ness within an inherently uncontrollable environment, this comes at the cost of any meaningful interactions with other inhabitants of the environment.

While the controlled use of headphones and other technology was considered acceptable by some hikers, what Krause calls "electromechanical" sounds on the trail were a source of frustration for many, who wrote about the particular irritation of road and railway noise. For Krause (2013: 158–159), noise is "an acoustic event that clashes with expectation...when there are discontinuities between visual and aural content, aural and aural content, or visual and visual content, these breaks are usually received by us as various kinds of noise". The discontinuity between a visual environment consisting of dense woodlands and an aural environment that included the sounds of cars rushing along on an unseen road seems to have been particularly jarring for some people. Their eyes may have told them that they were in the wilderness, but their ears told them otherwise. Pickering and Rice (2017) explore the implications of noise as "sound out of place" as a reformulation of Mary Douglas's famous classification of dirt as "matter out of place" (2002 [1966]). They argue that sounds are not intrinsically noise, but become noise when they occur in a place that they are not supposed to be. Hikers who drove or were driven to the AT to start their thru-hike are unlikely to have paid attention to the sound of other traffic on the road while they were travelling on it, but the same sound permeating the woodland of the AT 'wilderness trail' was problematic for many of them.

Krause (2013) argues that our brains work hard to filter out noise so that we can process 'desirable' aural information, spending a great deal of mental energy in the process. Several other sources have documented the detrimental effects of noise on human health (for example, see Dumyahn and Pijanowski, 2011). In general, the sounds of nature (geophony, biophony) are hardly ever regarded as noise, which we almost always think of as human-generated (see Fisher, 1999). Indeed, Fisher (1999: 28) points out that pairs of almost indistinguishable sounds, like thunder and bombs, or a roaring cataract in the mountains and a jet engine, have entirely different effects on their listeners, based on the source of the sound and not its sonic qualities: one is produced by 'nature', the other human-generated; therefore one is listened to with pleasure, while the other is regarded as noise. Fisher continues by pointing out that our bias towards the sounds of nature can lead us to enjoy sounds that purely from their sonic qualities we might be expected to dislike. His examples are the croak of frogs, the howling of wolves, the "guttural call of the Secretary bird", and the "nasal grunting of the osprey" (1999: 29–30). Indeed, many hikers indicated having relished the perceived strangeness of the animal sounds that they heard on the trail.

> [B]ullfrogs make such odd noises. They're somewhere between a pluck on an old guitar with loose, worn strings, and the weird sound my stomach makes.

While bullfrogs, as well as grouse, wild turkeys, cicadas, and other animals, were described as making strange sounds, nobody described these sounds as 'noise', and bloggers often got very creative in finding the right words to try to describe the alluringly odd sounds that they were hearing. The strangeness of some animal sounds was an intimate reminder of their Otherness, and was demonstrably enjoyed by many who wrote about them, in part specifically *because of* their strange Otherness.

Listening as witnessing

Chapter 1 touched upon the idea of witnessing in the context of pilgrimage. Paying close attention to the activities of animals on the trail can be seen as part of the 'witnessing' aspect of a nature pilgrimage, and hikers talked about their pleasure at being able to observe the activities of animals. Many of their narratives gave a sense that they enjoyed feelings of communion with other beings, as well as marginality in the presence of something 'bigger' than themselves, namely, 'nature'. Outside of the notion of pilgrimage, witnessing is still a useful concept to think about, particularly for those hiker-bloggers who paid close attention to animals on the trail and wrote in detail about what they had observed. Narratives about listening to animals on the trail convey a sense that animal sounds, both individual and collective, were frequently experienced by hikers who felt delight at being able to 'witness' animal lives through the sounds that they produced.

The 19th-century nature writer Henry David Thoreau paid close attention to the sounds of the natural environment around him, and wrote extensively about his experiences, for example about following the sound of a cricket and locating him underneath a rock, or the multiplicity of sounds in a woodland where he sat waiting at an owl's nest for her to make an appearance (Titon, 2015). Thoreau, who thought of nature sounds as the 'original' music, was very much a witness to the animals whose sounds he listened carefully to and wrote about. Titon describes Thoreau's appreciation for the voices of animals:

> Thoreau helps us understand that sound waves vibrate living beings into bodily experience of the presence of other beings. When that experience

and awareness is mutual, sounds vibrate beings into co-presence with one another. Sounds vibrate living beings into a way of knowing that proceeds by interconnection, a community of relations: a relational epistemology.

(Titon, 2015: 145)

From Titon's description of Thoreau's phenomenological concept, it is possible to think of the sound waves produced by trail animals as 'vibrating' hikers into an embodied experience of those animals. Perhaps this is what Schafer (1994: 11) means when he writes that "hearing is a way of touching at a distance". Titon argues that sounds contribute to a relational epistemology, making sounds not only a way of 'touching' (the animal) at a distance, but a way of 'knowing' (the animal) at a distance.

Fisher's proposal for why we value the sounds of nature over those that are human-generated, even when they share similar sonic qualities, is also helpful in thinking about this kind of 'sonic witnessing' of animals:

I propose that, being aware of [the natural sound's] origin, we hear it as a powerful and richly complex sound, and one caused by processes that, in being natural, are regarded by many of us as both *right* (they naturally belong) and *inevitable* – two aspects of the 'natural'... The sounds are made by creatures and processes that are themselves interconnected through evolutionary and geological processes. They are 'harmonious' in the sense that the things and processes making the sounds are harmoniously related to one another through their joint and interconnected evolution...Just as important, *soundscape* events in nature don't just belong together. They also belong *where* they are, they *belong to* the land. We hear them as belonging to their environment...

(Fisher, 1999: 35)

Fisher's explanation that nature sounds are satisfying to us because of our awareness that they have been created through evolutionary and geological processes speaks to the idea of people bearing witness to something bigger than themselves – the Other – which is very much what pilgrims hope to achieve.

He also emphasises the importance of place in the harmonious production of natural sounds. Place is of course very significant to the practice of pilgrimage, which centres around the notion of sacred space, and the traditional notion of witnessing is as a practice that occurs within sacred space. Whitehouse, too, argues that "the sounds of birds and other animals are importantly *sounds in place*" (2015: 58). He continues that, "for birds, sound-making is also place-making; it is an act of territorialising space, of making relations with other birds and continually re-weaving the context of their lives" (2015: 58). For hikers, witnessing birds and other animals "sound-making" did not just make them outside observers of acts of place-making; hikers engaged in sound-making themselves (through talking, breathing, and moving along the trail), and so can be seen as participants in the place-making activities of the inhabitants of the woods.

Listening and dwelling

[A]s place is sensed, senses are placed; as places make sense, senses make place.

(Feld, 1996: 91)

In a post entitled "The best and worst of long distance hiking", one hiker included in her list of the best things about hiking, "the music of the woods".

We heard so many wonderful sounds in the woods, some of which we never hear at home, like the barred owls and the whippoorwills. I loved falling asleep to the sounds of all the insects and frogs and birds in the woods.

Whitehouse (2015: 66) writes that "listening is not simply a process in which sound is heard but is a whole bodily experience of being in place in which sound is the focal point". When he talks about the farmer who feels a "lift" when listening to the dawn chorus every morning, he argues that the farmer's "sensing of place and time and of his own resonance with birds is grounded in his being-in-the-world" (2015: 64). The description of this farmer's sensory experience of "being-in-the-world" could also be termed 'dwelling', as in Ingold's "dwelling perspective" (2011).

In Chapter 2, which explored hiker experiences with bears on the trail, the dwelling perspective was helpful in thinking about how hikers and bears shared the space on and around the trail, each aware of the other, and adapting to the other's presence, while largely (although not always) avoiding contact. In other words, hikers and bears created an environment in which they were able – most of the time – to dwell alongside each other. Having looked at blogger narratives about hearing animals on the trail, it seems clear that animal sounds, both recognised and unrecognised, both miles and inches away, continually brought into being and recreated the environment of the AT for thru-hikers, as they, through the sounds that they made, contributed to the production of the environment for other animals.

For Ingold (2007), Schafer's term "soundscape" to describe a sonic environment is misplaced, ostensibly because it implies that the environment as "soundscape" can be carved out, extracted from the living world around us, and studied. Other commentators have similarly critiqued Schafer's conceptualisation of soundscape (see Feld, 2015; Helmreich, 2010; Kelman, 2010). Instead, Ingold describes sounds as being neither mental, nor material, but as a phenomenon of the experience of being immersed in the world. This sense of immersion was clearly felt by some, for example the veteran hiker who wrote poetically about "the sound of the woods" that "penetrates and saturates the human psyche", and of having been "absorbed in this aural environment".

Hearing ponies whinnying through thick fog, following the calls of a wild turkey, falling asleep to the sounds of crickets and cicadas, being newly able to recognise the rumble of a grouse taking flight, and having your laughter interrupted by a coyote howling in the distance were all embodied experiences of dwelling on the AT. In the end, dwelling is always dwelling *with*.

Earlier on this chapter quoted a blogger who wrote about what she termed the "horrifying" sounds of cicadas in the night, listing "layers" of other sounds, from something that sounded like a "faraway alarm clock" to trains passing by, and up to the sound of her own heart thumping. Indeed, a hiker's awareness of the sounds closest to her, the incidental sounds of her own body (breathing, footfalls, heart beats), as forming part of the sonic environment that she was immersed in can be viewed as proof of her absorption in the environment: she is a part of the environment, not apart from it. Yet at the same time, what hikers could hear in the environment often covered far more distance than what they could see around themselves, immersing them in a vaster and more expansive environment than their non-auditory senses.

Conclusion

Hiker narratives consistently described embodied and somatic experiences of being with self-willed animals, while at the same time never fully moving away from the notion of the wild animal as 'meaning something'. Yet hearing animals on the trail can arguably function as a way of recognising and attending intimately to their Otherness. Animal traces can have the effect of surprising and engaging those who come across them, as Hinchcliffe et al. (2005) found when looking out for water voles. Although they didn't see voles themselves, they learned to search for traces, such as footprints, and to use their noses in recognising the scent of water vole droppings. In the process they found that they could be surprised by water voles. They write that "our eyes (and to a lesser extent our noses) were being trained to recognise distinctions that were formerly invisible to us. The pictures, field signs and conversations were changing the way we sensed and…the way water voles made sense" (2005: 648). Water voles became interesting to them because the traces that they left spoke of the unexpected, much as the sounds of trail animals surprised and intrigued the hikers who wrote about them.

This is not to say that sound is not used in the construction of animal mythologies; it certainly can be, as DeLuca found, in his research on people listening to wolves, in which he points out that while the *Oxford English Dictionary* describes a howl as a "mournful cry", a wolf howl is not melancholic, but is produced in order to socially bond, rally, or mark territory (2016).

> [T]he howling message of a wolf is both a material object and a socially constructed metaphor that is infinitely interpretable and ideologically malleable, ultimately depending on the hearer's own values and biases.
>
> (DeLuca, 2016: 87)

AT hiker narratives provide a relational epistemology of sound – an acoustemology of radical Otherness (see Feld, 2015). For hikers, sound was another way of getting to know the nonhumans on the trail, and the *main* way of getting to know certain species, such as birds and insects. This despite the fact that most of the time hikers *didn't know* the intention behind, or reason for, the sounds produced by animals. What they could not have avoided recognising, however, was the autonomy behind the voices of self-willed animals in the woods.

References

Abram, D. 1994. Scattered Notes on the Relation Between Language and the Land. In: D. Clarke Burns (ed.) *Place of the Wild*, pp. 119–130. Washington, DC: Island Press.

Bruyninckx, J. 2011. Sound Sterile: Making Scientific Field Recordings in Ornithology. In: T. Pinch and K. Bijsterveld (eds.) *The Oxford Handbook of Sound Studies*, pp. 127–150. New York: Oxford University Press.

Bull, M. 2000. *Sounding Out the City: Personal Stereos and the Management of Everyday Life*. London: Bloomsbury Academic.

Carles, J. L., Barrio, I. L. and de Lucio, J. V. 1999. Sound Influence on Landscape Values. *Landscape and Urban Planning* 43: 191–200.

Clarke, E. 2005. *Ways of Listening: An Ecological Approach to the Perception of Musical Meaning*. Oxford: Oxford University Press.

DeLuca, E. 2016. Wolf Listeners: An Introduction to the Acoustemological Politics and Poetics of Isle Royale National Park. *Leonardo Music Journal* 26: 87–90.

Dickey, J. 1996. Springer Mountain. In: D. Emblidge (ed.) *The Appalachian Trail Reader*, pp. 85–87. New York: Oxford University Press.

Douglas, M. 2002 [1966]. *Purity and Danger*. London: Routledge.

Drever, J. L. 2014. Ochlophonia Hong Kong SAR: Audition, Speech and Feedback from Within the Crowded Soundscape. *Sound, Noise and the Everyday: Soundscapes in China*. Aarhus University, Denmark (21–24 August 2014).

Dumyahn, S. L. and Pijanowski, B. C. 2011. Soundscape Conservation. *Landscape Ecology* 26: 1327–1344.

Feld, S. 1996. Waterfalls of Song: An Acoustemology of Place Resounding in Bosavi, Papua New Guinea. In: S. Feld and K. H. Basso (eds.) *Senses of Place*, pp. 91–135. Santa Fe: School of American Research Press/distributed by the University of Washington Press.

Feld, S. 2015. Acoustemology. In: D. Novak and M. Sakakeeny (eds.) *Keywords in Sound*, pp. 12–22. Durham and London: Duke University Press.

Fisher, J. A. 1998. What the Hills Are Alive with: In Defence of the Sounds of Nature. *The Journal of Aesthetics and Art Criticism* 56(2): 167–179.

Fisher, J. A. 1999. The Value of Natural Sounds. *Journal of Aesthetic Education* 33(3): 26–42.

Gell, A. 1995. The Language of the Forest: Landscape and Phonological Iconism in Umeda. In: E. Hirsch and M. O'Hanlon (eds.) *The Anthropology of Landscape: Perspectives on Place and Space*, pp. 236–240. Oxford: Clarendon Press.

Helmreich, S. 2010. Listening Against Soundscapes. *Anthropology News* December 2010: 10.

Hinchcliffe, S., Kearnes, M. B., Degen, M. and Whatmore, S. 2005. Urban Wild Things: A Cosmopolitical Experiment. *Environment and Planning D: Society and Space* 23: 643–658.

Ingold, T. 2007. Against Soundscape. In: A. Carlyle (ed.) *Autumn Leaves: Sound and the Environment in Artistic Practice*, pp. 10–13. Paris: Double Entendre.

Ingold, T. 2011. *Being Alive.* London: Routledge.

Kelman, A. 2010. Rethinking the Soundscape. *The Senses and Society* 5(2): 212–234.

Krause, B. 2013. *The Great Animal Orchestra.* London: Profile Books.

Matsinos, Y. G., Mazaris, A. D., Papadimitriou, K. D., Mniestris, A., Hatzigiannidis, G., Maioglou, D. and Pantis, J. D. 2008. Spatio-Temporal Variability in Human and Natural Sounds in a Rural Landscape. *Landscape Ecology* 23: 945–959.

NPS (National Park Service). 2017. Natural Sounds Program. Available at: www.nps.gov/subjects/sound/

Passchier-Vermeer, W. and Passchier, W. F. 2000. Noise Exposure and Public Health. *Environmental Health Perspectives* 108(S1): 123–131.

Pickering, H. and Rice, T. 2017. Noise as "Sound out of Place": Investigating the Links Between Mary Douglas' Work on Dirt and Sound Studies Research. *Journal of Sonic Studies* 14. Available at: www.researchcatalogue.net/view/374514/374515

Pilcher, E. J., Newman, P. and Manning, R. E. 2009. Understanding and Managing Experiential Aspects of Soundscapes at Muir Woods National Monument. *Environmental Management* 43: 425–435.

Rice, T. 2015. Listening. In: D. Novak and M. Sakakeeny (eds.) *Keywords in Sound*, pp. 99–111. Durham and London: Duke University Press.

Schafer, R. M. 1994. *The Soundscape: Our Sonic Environment and the Tuning of the World.* Rochester: Destiny Books.

Schieffelin, E. 2005 [1976]. *The Sorrow of the Lonely and the Burning of the Dancers.* London: Palgrave Macmillan.

Smith, M. M. 2011. The Garden in the Machine: Listening to Early American Industrialization. In: T. Pinch and K. Bijsterveld (eds.) *The Oxford Handbook of Sound Studies.* New York: Oxford University Press.

Staples, S. L. 1996. Human Response to Environmental Noise. *American Psychologist* 51: 143–150.

Titon, J. T. 2015. Thoreau's Ear. *Sound Studies* 1(1): 144–154.

Whitehouse, A. 2015. Listening to Birds in the Anthropocene: The Anxious Semiotics of Sound in a Human-Dominated World. *Environmental Humanities* 6: 53–71.

Williams, B. 1992. Must a Concern for the Environment be Centred on Human Beings? In: C. C. W. Taylor (ed.) *Ethics and the Environment*, n.p. Oxford: Oxford University Press.

Zylinksa, J. 2012. What If Foucault Had Had a Blog? In: K. B. Wurth (ed.) *Between Page and Screen: Remaking Literature Through Cinema and Cyberspace*, pp. 62–74. New York: Fordham University Press.

Conclusion

It is along paths...that people grow into a knowledge of the world around
them, and describe this world in the stories they tell.

(Ingold, 2007: 2)

Internet users have become accustomed to 'Web 2.0', an online environment
in which users are not just consumers, but producers of content (see Stinson,
2016.). "Wilderness 2.0" is the name coined by Stinson (2016) to describe how
wilderness is now produced, consumed, and experienced online. Likewise, it
could perhaps be suggested that 'Animals 2.0' is an apt name for describing how
nonhuman animals can now also be produced, consumed, and experienced
online. When AT thru-hikers wrote about the autonomous animals that they
met on the trail, they re-created the presence of those animals in a digital
lifeworld. Yet this online re-creation of the animal was not so different from
any narrative retelling of an animal, and served largely the same purposes: to
describe what happened while engaging in a form of anthropocentric "self-
telling" (Bruner, 2004: 696) or "impression management" (Goffman, 1959) that
influenced how the narrator chose to tell the story. At various times in hiker
narratives, the author could be suspected of using the animal as a tool for the
self-telling project, as a prop, a set piece, as proof of something the narrator
wanted to convey about themselves. As Malamud states: "we have become
habituated to overwriting authentic, natural animals with a script that amuses
or benefits or otherwise satisfies our cultural cravings" (2012: 3), a statement
that harks back to Barad's argument that discourse is not what is said, but is
what frames the parameters of what *can* be said (2003: 819). This means that
even an attempt at a faithful retelling of exactly what happened can never
be entirely successful. As Merleau-Ponty says, "the thinker only ever thinks
beginning from what he is" (2014 [1945]: lxxxiii). Locating the 'real animal' in
an animal encounter story, therefore, is simply not possible. Luckily, that was
not the purpose of this research. Hurn asserts that there has been a general
consensus amongst anthropologists (e.g. Tapper, 1994) that they "need not
concern themselves with what animals are 'really like', but rather should focus
on what our human informants think of them" (2012: 209). While the 'animal'
and 'ontological' turns in anthropology have challenged this traditional view

in some quarters (e.g. Hurn, 2012: 210–211; Kohn, 2007), animals remain an important "mirror" (Haraway, 1991: 21) for reflecting information about how humans think about the world (Hurn, 2012: 3–4).

Biosocial perception

The Introduction to this book quoted Ingold (1994: 12), who asks whether nonhuman animals can meaningfully exist for us only as far as they are able to "exemplify an ideal type constituted within the set of symbolic values making up the 'folk taxonomy'", or whether it is possible for us to perceive them "directly, by virtue of their immersion in an environment that is largely ours as well". The constructionist viewpoint, that our perception of animals will always inevitably begin with what we have already been told about them, is propounded by Franklin (2012: 62), who writes that "in viewing animals we cannot escape the cultural context in which that observation takes place. There can be no deep, primordial relationship underlying the zoological gaze since it must always be mediated by culture". In the case of long-distance hikers on the AT, the preceding chapters have shown that their embodied encounters with wildlife on the trail could be intimate, fleeting, tentative, bold, surprising, delighting, frightening, shocking, and at times even bewildering – meaning literally that people found themselves 'lost in the wilds'. These somatic, immersed experiences with other animals led hikers to some very direct impressions of who they were interacting with, or who they were living among in the woods; a mother grouse allowing hikers to chase her in order to lure them away from her young, a shelter mouse boldly ransacking a hiker's food store, a highland pony clearly communicating his lack of interest in being 'petted'.

At the same time, hikers never fully moved away from the idea of the animal as 'meaning something'. They set out on their hikes with culturally acquired folk taxonomies of species, and perpetuated existing mythologies about wild animals through the narratives that they posted as blogs to the Trek online community. Often these mythologies centred around the animal representing 'wildness' or 'wilderness', which proved an elusive but endlessly alluring concept to hiker-bloggers. At other times the animals of the AT were seen as hosts, witnesses, obstacles, pests, mortal threats, friends, fellow travellers, and cuties.

What the previous chapters have demonstrated, then, is that hikers did not see the animals that they met on the trail solely as representing a particular cultural construction of their species, or through a pure and direct experience that led to Franklin's rather acerbically phrased "deep primordial relationship", but that their experience of other animals was, as Ingold (2013) might term it, biosocial. Hikers saw animals symbolically as welcoming woodland hosts on their pilgrimages *and* as real subjects with lifeways of their own. Bears could be metonyms for an untamed wilderness *and* autonomous animals lured by the smell of macaroni and cheese cooking. Moose could be mystical creatures representing the unknowable depths of the forest *and* the animal whose excrement a hiker just stepped in. To answer Franklin, it is correct to say that hikers couldn't quite escape the cultural context in which they encountered animals, but it is *also* correct to say that they often related to those animals in embodied, direct, intersubjective ways.

Becoming interesting

Fullagar (2000: 59) writes that "the vastness of nature's difference is experienced intimately as a self-other relation". One thing that is abundantly clear from AT hiker blogs about trail animals is that many hikers were fascinated – even enraptured – by the differentness of the life all around them. Animals of all shapes and sizes, ways of moving, of communicating, of vocalising, crossed paths with human hikers, exposing them to an almost overwhelming amount of corporeal Otherness. This extreme (in comparison to what most hikers were used to) diversity of life proved a challenge to dualistic human/animal thinking, when on the trail the human hiker was just one among many species. The Otherness of animals on the trail, while sometimes frightening or discomfiting in the moment, seems always to have been fascinating to think and write about afterwards. Some hiker-bloggers didn't just find themselves immersed in animal Otherness, they indulged in it, subsequently writing in detail about the varieties of life they had found on the trail.

As mentioned in previous chapters, Milton (2002) and Despret (2011), amongst others, have emphasised the importance of the animal being seen as interesting in order to be seen as worthy of consideration. The intriguing Otherness of the multitude of animals living on and around the trail was just one of the ways in which animals became interesting to hikers. The fact that they could surprise people with unexpected appearances, sounds or behaviours made them interesting. To people more used to domesticated animals, their autonomy itself – their freedom to move around as they wanted – made them interesting. Milton (2002: 103) writes that "sacredness depends on non-communication, on things remaining hidden...", and the fact that at times animals chose to elude the hiker's gaze, to fly away, or disappear into the woods, made them interesting. Hearing animals without knowing who produced the sounds made them interesting. The extended length of time that thru-hikers were on the trail for – up to seven months – meant that they became part of the lifeworlds of trail animals, as trail animals became part of theirs. Becoming more like them (becoming wild) – through sleeping, eating, and toileting in the woods – also made them interesting, as hikers began to think of themselves as more closely related to the animals on the trail than to the humans off it. Hinchcliffe et al. (2005: 643) write that "wild things become more rather than less real as people learn to engage with them". Hikers engaged with animals because they found them interesting, and this attention opened up more ways of knowing trail animals.

The cost of a trail animal becoming interesting was that his or her life was often disrupted by the consumptive practices of some hikers, who wrote about watching, approaching, following, chasing, taking photographs and videos, attempting to touch and pick up or feed animals that they found intriguing. The consumption of nonhumans as symbols (see Peace, 2005) sometimes resulted in being literally consumed in the flesh, as happened to the snakes who were killed and eaten by hikers. These narratives demonstrate that becoming interesting does not always result in care for the animal, and sometimes even results in consumption of the animal.

AT thru-hikers became the temporary inhabitants of a diversely populated woodland path, on which nonhuman animals had a significant degree of authorship over their own lives. Some hiker-bloggers focused their writing on the towns that they visited, the hostels that they stayed in, the people they met and food that they ate, while others wrote at length about the animal persons that they encountered while on the trail. It is clear from the stories that these narrators tell that there were no homogenous ways of relating to nonhumans, no one way to grow "into a knowledge" of the animals around them, as Ingold's quote at the beginning of this chapter suggests (2016: 3). However, certain recurrent themes stood out in the discourse, including the fact that animals were often viewed through the lens of the traveller on a pilgrimage, that black bears in particular invoked strong feelings as the subjects of a particular wilderness mythology, that a multiplicity of different beings could be viewed as cute, that certain animal bodies and actions could elicit powerfully fearful or disgusted reactions, and that hikers could come to know animals explicitly through the sounds that they made. I have written at length about how all of these themes demonstrate that animals were interesting to hikers. Ultimately, they were interesting because they were active; they *did* things. They were helpful agents in hiker-bloggers' self-telling projects precisely because they did things that caused hikers to respond, and they responded to what hikers did. They were sometimes viewed as objects, but at other times as subjects (see Cassidy, 2002), co-creating moments of intersubjectivity (see Hurn, 2012: 125–138) that were subsequently re-experienced through the act of committing them to narratives. Over the weeks and months, these animals became protagonists in the lives of hikers, and vice versa. By committing their encounter stories to the internet hikers not only ensured that trail animals were interesting for the period of time that they were on the trail, but made them an enduring part of their autobiographical projects.

References

Barad, K. 2003. Posthumanist Performativity: Toward an Understanding of How Matter Comes to Matter. *Signs: Journal of Women in Culture & Society* 28(3): 801–831.

Bruner, J. 2004. Life as Narrative. *Social Research* 71(3): 691–710.

Cassidy, R. 2002. *The Sport of Kings: Kinship, Class and Thoroughbred Breeding in Newmarket.* Cambridge: Cambridge University Press.

Despret, V. 2011. Experimenting with Politics and Happiness – Through Sheep, Cows and Pigs. *Unruly Creatures: The Art and Politics of the Animal.* London Graduate

School, Kingston University and the Centre for Arts and Humanities Research, Natural History Museum, London (14th June 2011). Audio available at: http://backdoorbroadcasting.net/2011/06/vinciane-despret-experimenting-with-politics-and-happiness-through-sheep-cows-and-pigs/

Franklin, A. 2012. *Animals and Modern Cultures*. London: SAGE Publications Ltd.

Fullagar, S. 2000. Desiring Nature: Identity and Becoming in Narratives of Travel. *Cultural Values* 4(1): 58–76.

Goffman, E. 1959. *The Presentation of Self in Everyday Life*. New York: Doubleday.

Haraway, D. 1991. *Simians, Cyborgs and Women: The Reinvention of Nature*. New York: Routledge.

Hinchcliffe, S., Kearnes, M. B., Degen, M. and Whatmore, S. 2005. Urban Wild Things: A Cosmopolitical Experiment. *Environment and Planning D: Society and Space* 23: 643–658.

Hurn, S. 2012. *Humans and Other Animals*. London: Pluto Press.

Ingold, T. 1994. Preface to the Paperback Edition. In: T. Ingold (ed.) *What Is an Animal?*, pp. xix–xxiv. London: Routledge.

Ingold, T. 2007. *Lines*. New York: Routledge.

Ingold, T. 2013. Prospect. In: T. Ingold and G. Palsson (eds.) *Biosocial Becomings: Integrating Social and Biological Anthropology*, pp. 1–21. Cambridge: Cambridge University Press.

Kohn, E. 2007. How Dogs Dream: Amazonian Natures and the Politics of Trans-Species Engagement. *American Ethnologist* 34: 3–24.

Malamud, R. 2012. *An Introduction to Animals and Visual Culture*. Basingstoke: Palgrave Macmillan.

Merleau-Ponty, M. 2014 [1945]. *Phenomenology of Perception*. Oxon: Routledge.

Milton, K. 2002. *Loving Nature*. London: Routledge.

Milton, K. 2005. Anthropomorphism or Egomorphism? The Perception of Non-Human Persons by Human Ones. In: J. Knight (ed.) *Animals in Person: Cultural Perspectives on Human-Animal Intimacies*, pp. 255–271. Oxford: Berg.

Peace, A. 2005. Loving Leviathan: The Discourse of Whale-Watching in Australian Ecotourism. In: J. Knight (ed.) *Animals in Person*, pp. 191–210. New York: Berg.

Stinson, J. 2016. Re-Creating Wilderness 2.0: Or Betting Back to Work in a Virtual Nature. *Geoforum* 79: 174–187.

Tapper, R. 1994. Animality, Humanity, Morality, Society. In: T. Ingold (ed.) *What Is an Animal?*, pp.47–62. London: Unwin Hyman.

Index